A Corpus-Informed Investigation into Environmental Engineering English

European University Studies
Europäische Hochschulschriften
Publications Universitaires Européennes

Series XXI
Linguistics

Reihe XXI Série XXI
Linguistik
Linguistique

Vol./Bd. 326

PETER LANG
Frankfurt am Main · Berlin · Bern · Bruxelles · New York · Oxford · Wien

Sylvana Krauße

A Corpus-Informed Investigation into Environmental Engineering English

PETER LANG
Internationaler Verlag der Wissenschaften

Bibliographic Information published by the Deutsche Nationalbibliothek
The Deutsche Nationalbibliothek lists this publication in the Deutsche Nationalbibliografie; detailed bibliographic data is available in the internet at <http://www.d-nb.de>.

Zugl.: Jena, Univ., Diss., 2007

27
ISSN 0721-3352
ISBN 978-3-631-57728-8

www.peterlang.de

Table of Contents

1 Introduction

1.1 Setting the Scene

Scientific progress is driven by the ever-increasing demand for new technologies. One of the sectors in which scientific progress is made at a very fast rate is the field of environmental engineering. Growing populations and diminishing resources are speeding up the demand for technologies which clean up contamination of air, water and soil and recycle spent materials. Nowadays, in a globalizing world, scientific progress takes place within co-operational relationships between international partners: multinational teams work on joint projects, records of scientific findings are available to a worldwide community of scientists and engineers, and products and services are provided across country borders. A mandatory prerequisite for such an international exchange is a common means of communication, the specialist language of environmental engineering. As universities are increasingly taking into account that students do not only need a profound subject knowledge in their specialist field but also the so-called soft skills with foreign languages being one of them, it is of vital importance that specialist English courses at university level take care of such needs and provide language courses which focus on what the language graduates will need to function in their professional lives.

It goes without saying that time which can be spent on the acquisition of language skills is scarce in the engineering curriculum. Therefore, the language focused on in class needs to be carefully chosen according to the laws of economy of effort. Research has shown that language for specific purposes is not general language "with a few terms thrown in" (Bowker/Pearson 2002:26) but a special variety with its own inventory of lexical, syntactic, morphological and stylistic features. Being able to master a specific language thus means to know and to be able to apply the language conventions typical for communication in that domain.

Earlier work has concentrated on showing that a particular domain-specific language is different from General English (cf. Nelson 2001). I will take this fact for granted and focus exclusively on the description of the language used by environmental engineers. This is done with the clear goal of applying the findings in the preparation of teaching and learning processes as well as during teaching processes themselves. With this I aim to contribute to remedying the lack of suitable information from almost non-existent reference and teaching material for the field of environmental engineering English. This study can be understood as an applied contribution to one central tenet in corpus linguistics expressed in Sinclair's words as follows: "Phraseology is due to become central in the description of English for Specific Purposes" (1996a:83).

In addition, this thesis is firmly based on some other basic principles and categories of corpus linguistics. It proceeds from the predominance of the idiom principle over the open choice principle in language use (cf. Sinclair 1987 and 2.1.5) and thus gives priority to multiword items rather than single words. To this aim, I adopt the different levels at which multiword items manifest themselves: collocation, colligation, semantic preference and semantic prosody (cf. Sinclair 1996a) to the use in a domain-specific language. As it is only when one knows a multiword item's constituents at these four levels and their interrelationships that it becomes "available as a single choice on the paradigmatic axis" (Tognini-Bonelli 2002a:91).

1.2 The Method

In order to derive the information necessary for a description of a domain-specific language at the lexical and the morphological level, a corpus consisting of texts from the domain in question was created, called ENGICOR throughout this thesis. This corpus was then searched with the help of modern corpus analysis tools such as Wordsmith 4 or MonoConc Pro and on the basis of corpus linguistic methodology. In line with the central assumption of the importance of significant frequency of language items, a keyword analysis conducted of ENGICOR compared to the British National Corpus (BNC) delivered 500 central items which were considered as most relevant for environmental engineering English. These 500 keywords were then subjected to a more detailed analysis at collocational and colligational level as well as on the level of semantic preference and semantic prosody hidden in the immediate co-text of the language item under investigation.

As a record of the information derived from this analysis has to be kept for the application in teaching processes, I developed the format of usage profiles which display concordance information in a more accessible way. On top of collocational and colligational information, these usage profiles were also designed to keep a record of morphological data. Part of the analysis was also a further categorisation of the most vital 500 keywords. Here again use was made of the corpus analysis tools in order to determine language items which are generally used over a range of texts from the domain in contrast to language items which are more specific for a certain subtopic of environmental engineering English. The investigation of the keywords again was undertaken with one superior aim in mind: the use in the language class.

1.3 Aims of Research

The motivation for the research undertaken within this thesis stems from a perceived gap of information on firstly what to teach in a language course specific to the needs of future environmental engineers and secondly how to teach it. To this end, the study follows three overall aims:

at first a very practical one, which is the creation of a bank of information on multiword items in environmental engineering English based on the exploitation of a small specialist corpus,

the second is a methodological one: establishing a method for corpus exploitation in order to gain relevant linguistic insights which should serve as a basis for teaching and learning processes in the foreign language class,

and thirdly, a theoretical one, the above-mentioned information-gathering and – analysis phases are always accompanied by applying main corpus linguistic theories to the investigation of a specialist language and thus using a specialist corpus as a testbed for the usefulness and applicability of basic corpus linguistic principles in a domain-specific language.

The information which is hoped to be derived from these investigations serves one superordinate goal: its application for pedagogical ends, be it on a descriptive level for pre-classroom use or on a practical level for in-classroom use.

1.4 Thesis Content

The thesis comprises five chapters. The first chapter gives an overview of the theoretical background and principles in corpus linguistics on which the methodological thinking in this study is based. It covers common corpus linguistic tenets like the idiom principle and the role of collocation as well as specific shifts in linguistic thinking which paved the way for the present study.

The second chapter starts with an overview of small corpus investigations in whose tradition this study follows. It continues by discussing the role of needs analyses in the process of preparing specialist language classes, explains the rationale behind the questionnaire sent out to engineering companies and discusses its findings in relation to the design of the corpus. Then the design of ENGICOR is focused on. The chapter finishes with a short outline of what I understand by environmental engineering English and its role as a domain-specific language.

In chapter 3 the corpus analysis process is described in detail, starting with a view of the importance of wordlists, followed by an evaluation of the use of keyword comparisons and a discussion on representativity and thus the usefulness and general applicability of the found data. Furthermore, the use of the usage profiles is exemplified and subsequently the information on collocational behaviour of chosen language items evaluated. Chapter 3 ends with a look at a higher level of meaning relations between lexical items: that of networks of meaning.

Chapter 4 is dedicated to an investigation into the morphological behaviour of language items typical for environmental engineering English. It starts with an account of various taxonomisation attempts by other researchers. On this basis, an own taxonomisation was established with a view to illustrating all word-formation patterns which contribute to this domain-specific language. Specific attention is paid here to the most common processes which lent themselves to the investigation by corpus research techniques, such as compounding, derivational, combining and conversion processes. Minor processes of word formation are just exemplified by corpus examples. This chapter furthermore touches upon questions of the productivity of certain word-formation patterns which might yield more new words in the future of environmental engineering English.

Chapter 5 describes how the established corpus data can be applied in language teaching. The question of how corpus-derived information on a specialist language can be used for language learning is discussed in three fields: for pre-classroom, in-classroom and post-classroom use. The changing role of the teacher as well as the role of the corpus in providing domain-specific subject information is discussed and recalling the information from chapter 4, it is also looked at how information on word-formation patterns can be included in class and which importance this has.

Finally, in the summary the findings of this study will be evaluated against the aims formulated at the beginning of the thesis and perspectives for future applications are given.

At the very end, the appendices contain all 500 usage profiles established on the basis of the corpus data as well as the keyword list and information from the detailed consistency analysis.

2 Chapter 1 – The Object of Study

2.1 Corpus Linguistics

2.1.1 J.R. Firth

Investigating the origins of corpora and corpus linguistics is not a typical chicken-and-egg-type dilemma; some underlying principles on which modern corpus linguistics are based were clearly laid down already before the advent of computer corpora within the framework of J.R. Firth's contextual theory of meaning. For him, language in the form of texts is embedded in the "context of situation" which he regards as the "essential basis for the study of meaning". The context of situation comprises all immediate constituents of language interaction: the participants, the objects and the effects of that interaction (Firth in Palmer 1968:108 ff.). Beyond the level of context of situation he puts the cultural context to which the concerned language relates and below the context of situation – and of more interest for this study – there is the linguistic context, or - as it is called in more recent times - co-text, in which words and phrases are embedded.

Groundbreaking for his time, it was also Firth who claimed language use - that is to say performance not competence - is central for deducing linguistic principles from. Texts should only be analysed in context and with regard to their function. This leads him to claim that "(r)estricted languages (i.e. domain-specific languages, my explanation) function in situations or sets or series of situations proper to them" (Firth in Palmer 1968:112), as environmental engineering English does within the professional setting of environmental engineers communicating with each other on domain-related topics with the help of domain-specific types of interaction. He furthermore says that "(d)escriptive linguistics of the structural kind is at its best when applied to a well-defined limited type or form of a major language,..." (Firth in Palmer 1968:98) and thus focused on the well-suitedness of domain-specific languages to be the object of descriptive linguistic activities. As early as in the 1950s, he already recognized the necessity of learning languages for specific purposes in the just emerging European Union. Consequently he called for micro-glossaries and micro-grammars of restricted languages to be designed on the grounds of descriptive linguistics (cf. Firth in Palmer 1968:106).

Returning to the thread of corpus linguistics, it was also Firth who started assigning great importance to the notion of collocation when he said that it "is the study of key-words, pivotal words, leading words, by presenting them in the company they usually keep" (Firth in Palmer 1968:106) that should be focused on. "The company they usually keep" further underlines the importance of repeated events which will be looked at in this study.

2.1.2 Corpora and Computers

It was then the emergence of computers which made analyses of language use possible on a large scale. Information about language use was recorded in electronic collections of written and transcribed oral texts - the corpora - which record the linguistic features ready to be analysed. Specific computer programs made it possible to look for form-based features in particular, such as lexical or morphological particularities, but also investigations into syntax and semantics were undertaken. Today, even phonological and prosodic information can be kept in corpora as texts can now be combined with video sequences and phonetic transcription.

According to Tognini-Bonelli, corpus work as the main methodological means of corpus linguistics is an "empirical approach to the description of language use" and "operates within the framework of a contextual and functional theory of meaning" using "new technologies" (2001:2). Commenting on the direction in which corpus linguistics is moving she furthermore says that the "primary goal of CDL (corpus-driven linguistics) is to make exhaustive and explicit connections between the occurrence and distribution of language items in text, and the meaning created by the text." (Tognini-Bonelli 2001:177)

Having mentioned that corpora are collections of texts in which language use is recorded and stored on a computer leaves me to add one more characteristic of corpora: they are designed as samples of a specific language variety or subsection, in the case of ENGICOR - the corpus used in this study – samples of environmental engineering English-specific language. In one of the most recent definitions of a corpus, Sinclair unites all these ideas: "A corpus is a collection of pieces of language text in electronic format, selected according to external criteria to represent, as far as possible, a language or language variety as a source of data for linguistic research." (Sinclair 2004).

2.1.3 Corpus-Based and Corpus-Driven Approaches

Whereas corpus-based investigations into language only reproduce and validate already existing theories and ideas in the different fields of linguistics, corpus-driven investigation starts not from ready-made assumptions but from the actual data. This data consists of repeated language items which serves as the basis for formulating hypotheses. Tognini-Bonelli comments on that method as follows: "The theory has no independent existence from the evidence and the general methodological path is clear: observation leads to hypothesis leads to generalisation leads to unification in theoretical statement." (2001:84-85) She furthermore says that the corpus-driven approach "aims to derive linguistic categories systematically from the recurrent patterns and the frequency distributions that emerge from language in context" (2001:87), a clearly inductive method. A corpus-driven approach is what this study aims at; no preconceived assumptions will be formulated in advance.

Observations of such recurrent patterns shook received linguistic knowledge in many fields. This phenomenon is aptly described by Sinclair's when he says that "(l)anguage looks rather different when you look at a lot of it at once." (Sinclair 1995:100) It led to the understanding that it is not words which are the bearers of meaning but rather larger chunks of language which became the new units of meaning. The new units of meaning will be dealt with in detail in 2.1.6. In the meantime the purpose of the following paragraph is to serve as illustration of the other parameters in linguistic sciences that prepared the ground for corpus linguistics.

2.1.4 Shifts in Emphasis

At this point I aim to summarise linguistic changes and tendencies which contributed to the model of thinking on which many corpus-driven and corpus-based studies on the lexical and meaning level are based.

Some contributing developments have already been touched upon above. First of all the advances in computer technology made the processing of huge authentic linguistic data possible. This led to a change in methodology: it was no longer intuition which the researcher's insight relied on but observations of such real data. This phenomenon is suitably described by Tognini-Bonelli as "a qualitative change in language description" (2000).

Scientific and technological progress also influenced linguistics and even language teaching in another way: with more and more specialised fields of work and knowledge exchange between experts and between different domains across boarders and cultures, the importance of mastering domain-specific languages has grown (and is still growing) rapidly. Linguistic research and teaching methodology based on this research has to meet this demand. As mentioned above, this thinking has its origin in the 1950s, when for instance Firth said about restricted languages that "(i)t is a more helpful approach than to set out and learn English in general – whatever that may mean." (Firth in Palmer 1968:106)

Whereas dealing with English for Specific Purposes meant at times just doing terminology work, there is now the conviction that it is phraseology which is of paramount importance (cf. Sinclair 1996a). Phraseology again, like Nuccorini rightly puts it, "is rooted in language use" (2002:10). The strong interest in formulaic aspects of language started amongst others with the work of Pawley and Syder on nativelike selection and nativelike fluency. Their research led them to postulate that "(t)he number of memorised complete clauses and sentences known to the mature English speaker is probably many thousands." (1983:205) Also Nattinger and DeCarrico's work on lexical phrases belongs nowadays to the standard research reports which triggered this change for a reinforced interest in formulaic language. Putting lexical phrases into a new perspective they claim that "(m)any early researchers thought these prefabricated chunks were distinct and somewhat peripheral to the main body of language, but more recent research puts this formulaic speech at the very center of language acquisition..." (Nattinger/DeCarrico 1992:XV).

Since then much linguistic research and many language teaching initiatives have been centred around the idea that meaning does not constitute itself in single words but in longer phrases. In the linguistic field, it is for instance Sinclair's idiom principle (see 2.1.5) that reflects that thinking. The most tangible reaction in teaching approaches resulted in the lexical approach put forward by Lewis (1993, 1997, 2000) and Willis (1990).

Sinclair's idiom principle opposes the long-standing slot filler model which goes back to Chomskyan times when words were only thought to fill the slots left by grammatically well-formed constructions. This together with the notion of colligation redefined the relation between lexis and grammar anew. They are no longer treated as two separate systems but as lexicogrammar where lexis is assigned much more power in constituting meaning and structures.

The change in attitude towards lexis has also had an impact on lexicography (cf. Teubert 1999). Lexicography based on corpus-linguistic principles calls for more linguistic contextual information which is reflected for instance in modern collocational dictionaries like The BBI Dictionary of English Word Combinations (Benson et al. 1997) or the Dictionary of Selected Collocations (Hill/Lewis 1999) for General English, and most remarkably for the German-speaking market of domain-specific dictionaries Fachwortschatz Medizin and Fachwortschatz Zahnmedizin (Friedbichler/Friedbichler 2003 and 2005).

In the pedagogical field new corpus-linguistic insights do not exclusively influence teaching in the way that learners will be compelled to concentrate more on formulaic language but also changes the way language is taught. The use of corpora and corpus-linguistics methods might counterbalance the tendency of the widespread communicative approach to focus too much on just exposing learners to the relevant language items. When being enabled to use corpora for information retrieval and checking of hunches, corpora become a powerful tool in the hands of learners which helps to achieve learner autonomy.

2.1.5 Open Choice Principle and Idiom Principle

According to Sinclair (1987), text – no matter whether it be spoken or written – is organised along two principles. Within the open choice principle, he explains, "at each point where a unit is completed (a word or a phrase or a clause), a large range of choice opens up, and the only restraint is grammaticalness" (Sinclair 1987:319-320). This concept gets close to the idea in generative grammar that every speaker can produce an unlimited amount of sentences and he or she is only restricted by the constraints of grammar. But only until the idiom principle comes into play which – as studies have shown – is the prevailing one. It says that " a language user has available to him or her a large number of semi-preconstructed phrases that constitute single choices, even though they might appear to be analysable into segments" (Sinclair 1987:320). Erman and Warren for instance investigated to which extent texts consist of prefabricated or formulaic language. They decided "that for anything to be a prefab the choice of one word must determine or, at least definitely restrict, the choice of at least one other, normally adjacent word" (Erman/Warren 2000:32) and came up with the claim that more than half of a text usually consists of prefabricated language. To illustrate and to reproduce this phenomenon, I chose a short piece of text from ENGICOR and looked for prefabricated or so to say phrasal items which I underlined to distinguish them from non-phrasal bits.

Fig. 1: Text sample from ENGICOR

Conclusions:
<Constructed wetlands> <have a great potential for> <contaminated wastewater treatment>. With careful design and planning, **<a constructed wetland> <can efficiently remove> <a variety of contaminants>**. **<The cost for>** design, construction and implementation **<can be considerably lower than>** other **<wastewater treatment options>**. **<The following set of> <case studies>** demonstrates **<the range of>** sites where **<this technology has been successfully applied>**. **<Groundwater contaminated with> <residual explosives> <was treated using> <constructed wetlands>**. **<A subsurface flow wetland>** and **<a surface flow wetland>** were constructed and compared. **<The subsurface wetland> <demonstrated a higher potential for> <contaminant removal>**.

I want to emphasise at this stage that I do not aim to be representative for the whole corpus but only to demonstrate how a text in relation to a corpus (and in relation to the usage profile) can display formulaic language.

17 out of 93 words from this little piece of text could not be assigned to any phrasal element, so 82% of it seems to be formulaic when phrasal elements were determined by their characteristic nature to occur in this form as repeated events in the corpus. The following table shows the phrasal elements and how often they occur in the corpus.

Tab. 1: Phrasal elements

sentence from the text	in usage profiles	in ENGICOR
Constructed wetlands have a great potential for contaminated wastewater treatment.	• -constructed wetlands • have great potential • potential for • contaminated wastewater • wastewater treatment	• constructed wetland* 79x • have * potential 53x • potential for 413x • contaminated wastewater 3x • wastewater treatment 502x
With careful design and planning, a constructed wetland can efficiently remove a variety of contaminants.	• constructed wetland • can efficiently remove	• constructed wetland* (see above) • * effectively remove 7x • a variety of contaminants 9x
The cost for design, construction and implementation can be considerably lower than other wastewater treatment options.	• wastewater treatment • treatment options	• the cost* for 51x • * considerably *er/more * than 18x • wastewater treatment (see above) • treatment option* 38x
The following set of case studies demonstrates the range of sites where this technology has been successfully applied.		• a/the set of 138x • case study/ies 227x • stud* demonstrat* 15x • range of 567x • technolog* appl* 5x • successfully applied • successfully applied 9x
Groundwater contaminated with residual explosives was treated using constructed wetlands.		• groundwater contaminated with 16x • contaminated with 139x • residual explosives 4x • * treated using * 31x • constructed wetland* (see above) **
A subsurface flow wetland and a surface flow wetland were constructed and compared.	• subsurface flow wetland • surface flow wetland	• subsurface flow wetland* 7x • surface flow wetland* 6x
The subsurface wetland demonstrated a higher potential for contaminant re-	• subsurface (flow) wetland • contaminant removal	• subsurface (flow) wetland* 7x

moval.		• demonstrat* * potential* 10x • potential for (see above) • contaminant removal 68x

**- * *treated using* * is part of a bigger pattern which can be expressed as verb+*ed using*, it occurs 581x in ENGICOR (*determined using, achieved using, calculated using, conducted using, measured using, performed using, processed using*, etc.), the N-1 position is occupied by objects, materials, parameters, activities and processes, the N+1 position is occupied by tools, agents, methods and technologies)

With the help of this table it can clearly be seen, how "whole clauses are made up of interlocking collocations" (Hoey 2005:5), in other words, how prefabricated language constitutes itself in text.

2.1.6 Collocation as Unit of Meaning

In the previous paragraphs, we were getting closer to the object of this study, the actual lexical items which occur with a certain frequency in the field of environmental engineering English. It has become clear that the object of study cannot be a single word or term but a multiword item as the meaning load of ideas is usually spread over several lexical items. Research into cognitive encoding and decoding processes have shown that prefabricated chunks of language demand less cognitive energy when uttering text as well as when understanding text (cf. Pawley and Syder 1983) and this is how human communication works effectively. Moreover, as prefabricated chunks account for such a high percentage in text it is, to say it in Stubbs' words, "increasingly argued that collocation is a fundamental organizing principle of language in use." (Stubbs 2001:60).

The central statement that most definitions on collocation make is that they represent the co-occurrence of at least two lexical items. This is sometimes seen strictly statistically, so that word combinations like *but also* or *tailings from* would count as a collocation. As this makes no sense from a language learner's point of view, I would like to add that collocation needs to be meaningful, that is to say only semantically and syntactically complete items are considered here as collocations. For a workable definition, I chose Wray's definition of collocation as it provides the relation to the cognitive processes as well which are of great importance for language learning. According to Wray, a collocation is " a sequence, continuous or discontinuous, of words or other elements, which is, or appears to be, prefabricated: that is, stored and retrieved whole from memory at the time of use, rather than being subject to generation or analysis by the language grammar." (2002:9).

A look at the usage profiles of the 500 keywords is enough to see that collocation most frequently comes in the form of the co-occurrence of two lexical items, 3-word or 4-word and even more so 5-word items are rarer but are still counted as collocations (i.e. *uranium mill tailings disposal site*). To be chosen for inclusion in the usage profiles, recurrent lexical items had to occur at least 5 times in the corpus.

I will also use the term "multiword items" in this study which designates collocations as well as recurrent patterns of items which go beyond collocations. In items like for instance *drainage from underground workings*, there is the collocation *underground workings* and the colligation *drainage from* but as these words occur frequently together, the whole phrase is considered a multiword item. There was no specific collocational span chosen; the natural boundaries for collocations were assumed to be the sentence boundaries.

Multiword items were identified in this study using the concordancing tool of WordSmith4 or MonoConc Pro. Concordances of search terms were then scanned by the observer for suitable language items. This "eyeballing" technique allows one to look for not so obvious patterns as for instance in *plume of chlorinated hydrocarbons/chlorinated solvents/chromate and TCE/contaminants/contaminated water/gasoline/*etc. which only reveals its pattern of *plume of any kind of contaminant* to the human observer, not to the concordance tool.

2.1.7 Extending the Notion of Units of Meaning

In the preceding paragraph, a suitable definition of collocation was decided upon and underlined how important it is for the language learner to know which lexical company a word keeps. But successful language learning cannot stop at this level. Sinclair argues in his model of extended units of meaning, that in addition to the collocational knowledge, it is also the knowledge of a word's typical colligational behaviour and the semantic preference and semantic prosody it displays. This makes the lexical item complete in terms of using it successfully in the right linguistic context and in the context of situation (cf. Sinclair 1996a).

If we take the example of *exposure*, it means that students need to know the collocational profile of this word like *permissible* or *prolonged exposure, radiation/sunlight/UV exposure, human exposure, exposure assessment/ concentrations/conditions/levels/pathways/route/time* as well as the colligational profile of *exposure to carbon hydroxide, dust/pathogens/TCE/to UV radiation/...* together with the semantic preference for a sort of radiation which is revealed by the semantic set of *light/sunlight/radiation/UV (exposure)* and the semantic prosody of the damaging influence of such radiation.

These four steps in the identification process of a lexical item, or so-called extended unit of meaning, are also retraceable in the usage profiles (see 4.2 and Appendix B) which portray the 500 most frequent keywords of environmental engineering English. Collocation, colligation, semantic preference and semantic prosody are the four components which are considered central for the successful learning of a (domain-specific) language in this study.

In his "new theory of language" Hoey builds on Sinclair's extended unit of meaning and applies it to more text-centred categories (cf. Hoey 2005).

2.1.8 Hoey's New Theory of Language

Hoey claims very much like Sinclair that lexical items are primed for collocational use. By primed, he means, that "As a word is acquired through encounters with it in speech and writing, it becomes cumulatively loaded with the contexts and co-texts in which it is encountered, and our knowledge of it includes the fact that it co-occurs with certain other words in certain kinds of contexts." (Hoey 2005:8).

In addition to collocational priming, he continues to say that words are also primed for semantic association in that a word can be associated with a semantic set or class it collocates with. Here we can find the idea of semantic preference mentioned above just as we can find the concept of colligation when Hoey postulates that lexical items are also primed for colligation, i.e. the grammatical functions or structures in which a word occurs (cf. Hoey 2005:38 ff.) and the concept of semantic prosody which Hoey gets near when he says that lexical items are also primed for pragmatic associations.

In his claim that a lexical item also has grammatical priming which maintains that words are typically primed for use as a noun, verb, adjective, etc., he looks at word class and the immediate co-text of words. But Hoey also looks further than the collocational span within the single sentence and moves to text level when he says that lexical items are furthermore primed for specific textual colligation, that is the preferred textual position of that item, for instance at the beginning of sentences, paragraphs or texts. It can also be primed for textual collocation, i.e. primed for being cohesive as there might be words in the text which take up the ideas mentioned before in form for instance of a pronoun which stands for a certain object. And finally there is the idea of textual semantic association, a notion which might remind us about the aboutness of text as it says that talking of a specific topic, there are a few words very likely to occur when the context of situation is alike (cf. Hoey 2005:115).

What is furthermore worth noticing in Hoey's model is, that he is convinced that collocational priming is also sensitive to the domain in which the lexical item is encountered (Hoey 2005:10). That is to say the same item can have different collocational environments in different domains, like for instance the word *pregnant* in environmental engineering English which preferably collocates with *liquor* and *solution* rather than with *woman* in General English.

The consequences of both, Sinclair's extendend unit of meaning and Hoey's priming model is that lexis again is assigned the most important role in language learning, with grammar losing more and more ground. Teachers consequently have to make intensive use of authentic language materials which provide many encounters with the relevant lexical items to speed up priming (Hoey 2005:183 ff.).

Having a database of 500 analysed words at hand, one could ask the question as to whether collocational behaviour of certain lexical items resembles each other. This could be of importance for drawing conclusions during language learning processes. To this end, I am asking whether generalisations can be made of items which play a similar role within certain topics, i.e. recyclable materials. Can I assume that items like recyclable paper or recyclable plastic can be found in similar collocational frames? This question will be further examined in 4.3.2.

2.1.9 Colligation

Referring to Firth mentioning the notion of colligation at a very early point in time, Tognini-Bonelli sums up the current meaning of colligation as follows: "In the light of corpus evidence, however, the concept of colligation is being extended to include the syntactic constraints, or indeed preferences, that a specific word, seen as a unique lexical item rather than as a member of its class, entertains within its environment" (2001:89). According to this definition, colligation manifests itself for example in a word co-occurring with a particular preposition (i.e. *adsorption (on)to aquifer sediments*) or the obligatory use of the *–ing* form of a verb (i.e. *in addition to *ing*).

Whereas it was found that collocational behaviour is to a very high degree arbitrary, at least some colligations were felt to be part of higher patterns (i.e. *sorted by* follows the pattern of passive constructions indicating the agent or method of the action, or *at chemical equilibrium* can be seen as belonging to a similar pattern as *at a concentration of ..., at constant rate, at a pH of..., at depths of..., at maximum efficiency, at room temperature, at steady state*, all describing circumstances under which reactions take place starting with the preposition *at*).

Colligations can be found in the usage profiles under the clusters section, as the multiword items resulting from them are usually longer than two words. As in the following example of *plume of*, the subsequent, more varied part of the multiword item is then only illustrated by some examples which usually give a good idea which semantic information it reveals on top of the colligational information.

That is, in the usage profile you would only find *plume of (28) chlorinated solvents/contaminants/dissolved halocarbons/PCE/TCE/...*

Fig. 2: Concordance of *plume of*

... ative rates of migration of a [[plume of]] **a colloid-mobilizing agent** a ...
... cludes the cost of cleaning a [[plume of]] **chlorinated solvents** using ...
... ed. As a result, there is a [[plume of]] **chlorinated solvents** located ...
... rolina, to intercept a mixed [[plume of]] **chromate and TCE**, following t ...
... the subsurface to intercept a [[plume of]] **contaminated ground water** w ...
... ed in the path of a migrating [[plume of]] **contaminated groundwater**. Re ...
... e groundwater below created a [[plume of]] **contaminated water** that, by 1 ...
... at the Site is Cr (VI). The [[plume of]] **Cr (VI)-contaminated groundwa** ...
... 1987. In 1990, a 1200 x 190 m [[plume of]] **dissolved chlorinated solvent** ...
... ue to generate a continuous [[plume of]] **dissolved halocarbons** (Natio ...
... er, when dealing with a large [[plume of]] **dissolved product** there is a ...
... t to contain and remediate a [[plume of]] **dissolved TCE** located in a sh ...
... ydraulic barrier to contain a [[plume of]] **gasoline and diesel fuel** in ...
... s installed in the path of a [[plume of]] **groundwater containing Cd**, Cu ...
... 95). This system intercepts a [[plume of]] **nitratebearing ground water** ...
... mical difference imposed by a [[plume of]] **secondarily-treated sewage**; r ...
... sulted from the presence of a [[plume of]] **sewage-contaminated groundwa** ...
... rees to control and degrade a [[plume of]] **TCE** in a shallow aquifer. Th ...

2.1.10 Semantic Preference and Semantic Prosody

Semantic preference and semantic prosody are the two linguistic features that according to Sinclair complete the notion of a unit of meaning and are on a more abstract level than the other two features collocation and colligation discussed earlier on. In Hoey's terminology, certain words are primed for use with certain semantic associations and certain pragmatic associations (Hoey 2005:16ff and 26ff).

Semantic preference and semantic prosody together with collocational and colligational behaviour is considered by this study as the kind of lexical information a learner needs to acquire in order to successfully use a lexical item in discourse. There is not much disagreement in the literature on the notion of semantic preference. Stubbs (2001:65) defines it as "the relation, not between individual words, but between a lemma or word-form and a set of semantically related words". A familiar example of such behaviour is the common textbook example of verbs indicating shifts in trends like *decrease, increase, drop, rise* which preferably collocate with qualifying adverbs like *slightly, dramatically, significantly, gradually, slowly, steadily*, etc.

The concept of semantic prosody though seems to be not that easily to grasp and fosters quite some discussion on whether it is a constituent of every lexical item or not as well as by which means it gets expressed. Partington defines semantic prosody as "the spreading of connotational colouring beyond single word boundaries" (1998:68). And for Sinclair (1999) semantic prosody develops "When the usage of a word gives an impression of an attitudinal or pragmatic meaning,..." The words "connotational" and "attitudinal" from these definitions trigger some serious intuitional doubts whether the concept of semantic prosody is applicable to such a sober and neutral thing as a specific language that is described on the basis of a corpus of scientific texts. This will be the object of investigation later on.

2.1.10.1 Semantic Preference in ENGICOR

What the intensive work with lexical items of a specific language shows is that "their range of collocates is (...) quite limited" (Rundell 2000). Following the idiom principle, the words a lexical item can co-occur with are not arbitrary. The intensity to which the individual lexical items are limited is variable. Some words like *include* collocate with such a wide variety of collocates that the endeavour to group them into semantic sets is futile. Others have clear semantic preferences which can be detected in their concordance lines and are consequently documented in the usage profiles. One of these words is for instance *mining* which collocates with the lexical sets of materials (as in *coal mining, copper mining, phosphate mining* or *uranium mining*), of specific technologies (as in *biomining, solution mining* or *hydraulic mining*) and of locations (as in *open-pit mining, surface mining* or *underground mining*).

A look at the usage profiles suggest that the phenomenon of semantic preference is much more frequent than common ESP-textbooks imply. Semantic sets with which certain lexical items collocate come in all shapes and sizes, i.e.:

- as the modifying noun of a compound that displays the semantic set of recyclable materials (i.e. *biosolids recycling, electronics recycling, packaging recycling, paper recycling*, etc.),
- as the headword in a compound premodified by adjectives that collocates with certain waste substances (i.e. *biodegradable fluids, biodegradable organics, biodegradable slurry, biodegradable waste*, etc.),

- as an adjective that collocates with a set of graduating adverbs (i.e. *extremely acidic, highly acidic, strongly acidic, moderately acidic, slightly acidic, weakly acidic*)
- as part of an infinitive construction (i.e. more than half of the 914 occurrences of *designed to* are followed by an infinitive carrying the meaning of an action or a state - as in *designed to achieve, designed to capture, designed to carry, designed to contain, designed to control*, etc.)
- as part of a cluster which allows internal variation belonging to the semantic set of descriptions of process conditions (i.e. *under acidic conditions, under (un-)aerobic conditions, under controlled conditions, under fuel-rich conditions, under laboratory conditions, under reducing conditions*, etc.) or the semantic set of years and dates and the one of processes which collocate with prior (i.e. *prior to 1942/1999/etc.* or *prior to chlorination/discharge/disposal/filtration/injection/sampling/use*, etc.).

And this list is by no means exhaustive; many more examples of lexical items collocating with specific semantic sets are to be found in the usage profiles (see Appendix B).

2.1.10.2 Semantic Prosody in ENGICOR

Semantic prosody, however, is more difficult to detect and to describe. The question whether there is a phenomenon like semantic prosody in environmental engineering English can be answered with a definite yes. Examples found in the usage profiles will show this later. But whether all lexical items display a semantic prosody remains doubtful if one does not resort to using the notion of a neutral semantic prosody in case there can be no positive or negative prosody detected.

Five lexical items taken at random were examined in order to establish their semantic prosody. *COD:DOC ratio, analytical chemistry, mechanical properties, filter effluent* and *determine the volume* are all clearly phrases from environmental engineering English and certainly display no semantic prosody at first sight. Nevertheless, widening the view a bit with the help of the concordance lines, two of the five seem to have a positive prosody as can be detected in Figure 3 and 4.

Fig. 3: Concordance of *analytical chemistry*

the manufacture of matches and in [[analytical chemistry]], but more than 90 percent of
the team is collectively skilled in [[analytical chemistry]], environmental engineering, st
ge and skills, including expertise in [[analytical chemistry]], environmental engineering,

ding of this document, expertise in [[analytical chemistry]] and statistics are necessary fo
ysis of hazardous organic emissions, [[Analytical Chemistry]], 54, 810, 1982. (b) Use th

Fig. 4: Concordance of *mechanical properties*

lmethacrylate and found that enhanced [[mechanical properties]] were possible. In contr
rated during this study for enhancing [[mechanical properties]]. Of the two techniques, th
ated during this study for enhancing [[mechanical properties]]. A peroxide reaction was u
e resulting materials show excellent [[mechanical properties]] and may be suitable for a
elastomers were observed to improve [[mechanical properties]] of the resultant blends; h

The number of lines is quite low, but what I meant to show with these – at first sight rather unsuspicious examples- is what Sinclair, speaking about semantic prosody, expressed with the following words: "if you dig deep enough it is there" (personal communication). With the other items, however, no semantic prosody could be detected with the help of the investigated corpus which makes me also quote the end of Sinclair's phrase: "if only it is a neutral one". See Figure 5 for the concordance of *filter effluent* where ENGICOR could not reveal any kind of positive or negative prosody. Neither could this be found for the remaining two examples.

Fig. 5: Concordance of *filter effluent*

ory analysis were collected monthly. [[Filter influent]] and effluent samples were collect ng
th the waste strength data for the [[filter influent]] and effluent. The information in S
ent system. 2.6.1 Sample Collection [[Filter influent]] and effluent samples were collect
one year. Laboratory analysis of the [[filter influent]] and effluent showed an average re

A different view gets is brought into play by Partington who says that "One view would be that semantic prosody is a sub-category, or a special case, of semantic preference" (2004:149). That means that in addition to a significant co-occurrence of a lexical item with a semantic set, this semantic set displays a certain positive or negative semantic prosody. A classic and oft-cited example is the verb *cause* which is followed in 90% of the cases by unpleasant circumstances like *accident*, *crisis* or *difficulty* (cf. Stubbs 1997). ENGICOR confirms Stubbs findings this time for the domain-specific language: the most frequent collocates of *cause* are *clogging, damage, fouling, harm, methemoglobinemia* and *problems*.

Having accepted to work with the notion of semantic prosody, the next question would be to ask where or by which means it gets expressed in the lexical items. The answer to this question will be the key for the language teacher and learner to look for this phenomenon.

By the help of skimming through the usage profiles, several more examples of semantic prosodies were found. At the most tangible level, it was found that some lexical items come with a semantic prosody that lies within their own central meaning load and is thus accessible by the intuition of the language users. Words that belong to this category are *hazardous, difficult, valuable* and *effective*. A second – at first sight rather neutral group of words – acquire semantic prosody in language use that is in the immediate linguistic co-text they appear in. Examples for such a behaviour are for instance: *potential* as it occurs in *neutralization potential, methane-recovery potential* displaying a positive prosody and a negative in *potential for acid generation, potential for problems, potential to cause corrosion or AMD.*

Morphological components of lexical items are also found to provide room for semantic prosody.

All words starting with the neoclassical combining element *bio-* seem to have a positive prosody around them which stems from their frequent use in positive contexts: their priming or conditioning in other words. This phenomenon was either consciously or unconsciously made use of when choosing the term *biosolids* to designate *sewage sludge* (compare 5.2.10). Moreover, on grounds of their central meaning load a suffix like *–able* also seems to have a positive prosody around it, although I would not go as far as to claim that all morphological derivational elements are home to a specific prosody.

As in the example of *bio-*, the social context is also suspected to play a role in assigning or at least strengthening a semantic prosody in some lexical items like in *natural* and *artificial, innovative* and *conventional.*

Referring back to the definition of semantic prosody given in the previous paragraph, the examples from the corpus show one remarkable fact: instead of depending on speaker attitude or feelings, they are language-immanent. People do not choose to say *cause erosion/damage/clogging/fouling* or *minimize the risk/impacts/potential for/costs for/effects of* because of their attitude towards the expressed content of speech. They are simply expressing reality which can hardly be described in other words. Speaker attitude is expressed in sentences displaying what is traditionally called connotation, a term that can get mixed up with semantic prosody but which has to be kept apart, at least in the field of engineering language. To my knowledge, ENGICOR does not contain any example where a particular connotation is expressed. For this reason, I must reject the above-mentioned definitions of semantic prosody by Partington (1998) and Sinclair (1999) as evidence for domain-specific language shows that speaker attitude is not involved. A workable solution is to follow Partington in his later suggestion to call semantic prosody a subset of semantic preference where the semantic set which regularly occurs with a certain lexical item is to a high percentage of the examples additionally associated with either positivity or negativity (Although it has to be mentioned that Partington goes on at this stage to claim that the notion of semantic prosody is more complex than that and that in his view "semantic prosody is evaluative or attitudinal". His claims are made for general and academic English (cf. Partington 2004:150).

Rounding up, I want to show with the following example, how semantic prosody gets created in language context and how it can be made visible by concordancing. In order to illustrate an either positive or a negative semantic prosody which develops out of the interplay between a verb and its complement I chose the search term *significant* as a magnet to attract this construction, as *significant* serves in such cases as an intensifier. The examples which generate a negative semantic prosody are marked in grey, the positive ones remain white. Note the example of *result in significant* + complement which can either be positive or negative.

Fig. 6: Semantic prosody around *significant*

sal of the lime is **considered a** [[significant]] **problem** by the Washington State Departm
dicates that this is not **considered a** [[significant]] **risk**, as the models used are very con ...
the lime by-product **is considered a** [[significant]] **problem** by the Washington State Dep
.mbers of seagulls at the site **pose a** [[significant]] **threat** to aircraft from a nearby airp ...
tems 1. Abstract Scrap tires **pose a** [[significant]] **environmental and public health proble**
d gasoline/ fuel components, **pose a** [[significant]] **human health and environmental risk**
wastepaper as feedstock would **pose a** [[significant]] **burden**. In light of the responses in t

sses. The plastic processors **achieve** [[significant]] **reductions** as well. Their initiatives—
n necessary for the system to **achieve** [[significant]] **improvements** in environmental perf
for destruction, but also to **achieve** [[significant]] **energy or materials recovery**. The reg
ce Track, they had already **achieved** [[significant]] **reductions** in their discharges to water

ination of surface waters can **cause** [[significant]] **human health problems**, from simple
iants of ammonium nitrate can **cause** [[significant]] **loss of ammonium nitrate** from the s
ithout toxic constituents) can **cause** [[significant]] **adverse impacts** on stream quality and

eated Wastewater sludges also **contain** [[significant]] **quantities** of the three main elemen
r are often highly acidic and **contain** [[significant]] **quantities of suspended solids**. Under
influent wastewater does not **contain** [[significant]] **levels of settleable solids**, greases a ...
te or mordenite and does not **contain** [[significant]] **quantities of the minerals smectite** (m

ers high in carbonate might **result in** [[significant]] **buildup of calcite** ($CaCO_3$) or siderite
tural wetlands, which can **result in** [[significant]] **short circuiting of flow**. Dye tracer s ...
tely treated wastewater can **result in** [[significant]] **habitat and fisheries damage**. Currentl
.nce 3 % shingles does not **result in** [[significant]] **savings**, it is recommended that test se

e excellent performance and **result in** [[significant]] **savings** by reducing the amount of vi
winter months, was found to **result in** [[significant]] **net infiltration** (~30% of annual pr
n used effectively, it can **result in** [[significant]] **waste reduction and cost savings**. Pr ...
total oil and grease would **result in** [[significant]] **removal of these pollutants**. Covere ...
a lower redox potential can **provide** [[significant]] **removal** as was shown by Young et al.

.crap glass can in most cases **provide** [[significant]] **cost savings** to users.43 Is there dema
educed aeration demand could **provide** [[significant]] **cost savings** for a wastewater trea
ause living plants and litter **provide** [[significant]] **frictional resistance** to flow through ...

2.1.10.3 Semantic Preference and Semantic Prosody in the Learning Process

What is being claimed for collocations in terms of presenting and learning the items together with the company they keep holds true for semantic preference as well. Teaching and learning materials should go far beyond the common textbook example of *increase + steadily/dramatically/slightly*. Where there is a valid semantic preference, it should be made obvious to the language learners and that mastering of units of meaning are strengthened by repeated encounters with these items. Learners can only understand by authentic examples that it is a *plume of + any kind of contaminant*, a specific phrase which becomes imprecise and difficult to understand when expressed in other words. The same goes for semantic prosody. Where semantic preference and semantic prosody delimit the use of lexical items, learners need to pay attention when speaking as the infringement of the underlying rules result in misunderstandings. The way to acquire this knowledge is frequent exposure to authentic examples and conscious processing triggered by the teacher. Intuition and paraphrasing their way around semantic preference and semantic prosody are certainly the wrong techniques.

3 Chapter 2 - Towards a Corpus

3.1 Small Corpora Applications

A corpus in the modern sense is a collection of texts in computer-readable format ready to be exploited by corpus-search tools. The main methodological steps of corpus linguistics consist of the "automatic extraction of language data from corpora; output processing by mostly statistical procedures, and finally the validation and interpretation of this processed data" (Teubert 1999). The corpus has to be designed in such a way that actual instances of the desired language items can be observed by concordance software which displays all samples of instances in one diagram window.

Lynne Flowerdew comments on the current development of corpus research as follows: "Over the last decade, the field of corpus linguistics has widened in two senses, i.e. now encompasses more and bigger corpora and also small corpora." (2001:363). On the one hand there is the trend to create ever bigger corpora of general language (i.e. Bank of English had in 2005 525 million words or the British National Corpus with its 100 million words). On the other there is a vast number of small corpora of specialised language which are being created and investigated. What "big" means in such dimensions is obvious from the immense number of running words in corpora like the BNC or the Bank of English; "small", however, is relative depending on the text type selected: 50,000 words of for instance job adverts can constitute a viable corpus for exploring this genre, 50,000 words from technical reports not.

Like large corpora, small corpora are designed to serve as smaller-scale models of the examined language in use. They are created with the aim of delivering insights into the problems the corpus designers want to study. Sinclair (2001:XI) points out that large and small corpora differ not only in size but also in the research methodology applied to extract linguistic information. Small corpora are built for EHI – Early Human Intervention, meaning that useful observations can be made without much prior tagging and parsing which is often beyond the scope of the small-scale ventures of small-corpus designers. Frequency and keyword analyses and concordancing are the main corpus manipulations being applied which allow for comparisons in order to detect special features of the linguistic specialism investigated.

When Ma wrote his meticulous account on "Small Corpora Concordancing in ESL Teaching and Learning" in 1993, the advances in the use of computers for linguistic studies had brought about multiple attempts to use small corpora for what he categorises as (a) syllabus design, (b) classroom teaching and (c) test construction (1993:11). He was yet still able to give a comprehensive overview of the various developments going on at that time. Now, more than ten years after, small-corpus investigations have increased considerably. They all contribute by providing one piece of information on one particular domain-specific language, one particular text type. ENGICOR and the linguistic knowledge drawn from it constitutes one further piece in a big puzzle.

Coming back to Ma (1993), he gives the mid-80s as the starting point for when scholars like Johns and Higgins (Johns/Higgins 1984, Johns1986 and 1988) began using concordancing data in the classroom, blazing a trail for the Data-Driven Learning approach. He then mentions John Flowerdew (1991) as the one beginning to use corpus data to evaluate existing course programmes and course materials and to develop new ones based on corpus work (cf. Ma 1993:13). Within the context of corpora for teaching purposes, Ma furthermore speaks of the use of ESP learner corpora in order to detect "recurrent patterns of errors" (1993:17) made by language learners and the beginning use of parallel concordances to compare certain features across different languages (cf. 1993:22) and goes on to report on the first application of corpus information for test design (cf. 1993:23).

The different fields of application distinguished in Ma (1993) stayed the same showing tendencies to spread into other scientific fields like history or forensics. Yet what drove the first users of small corpus information was the fact that small specialized corpora allow for observation of lexis and the identification of typical lexical patterns. "(T)he identification of specific terminology is of paramount importance in LSP" as Tognini-Bonelli puts it rightly (2001:8). The specialized corpus is used as a shortcut to the target specialist language.

Most small corpus investigations reported in Ma (1993) and beyond get their thrust from a perceived knowledge gap. Dictionaries and teaching and learning materials are felt to either provide not enough information on certain linguistic features or even to a certain extent the wrong information. A drastic example of this are the specialist dictionaries which only provide one-to-one equivalence lists of specialist terminology and thus leave learners alone with the problem of how to put these words and phrases into the context of other linguistic items.

In order to bridge this gap, many investigations have been undertaken which mostly concentrate on establishing lexical and syntactic patterns which provide for a fuller picture than conventional dictionaries and teaching materials can offer. I will here mainly concentrate on the ones aiming at improving ESP teaching, both with the aim of improving course design and with direct application in the ESP classroom.

Researchers who can be seen in the tradition of aiming at the detection of linguistic patterns in domain-specific languages for subsequent use in teaching are for instance Curado (2001 and 2002) who extracted a common core vocabulary from his Business Science and Technology Corpus as well as Flowerdew who also aimed at extracting what he called subtechnical vocabulary in a specialist biology corpus. Other studies focus on the investigation of specialist vocabulary like Williams (1999) who concentrated on establishing a pedagogic dictionary based on the observations of his corpus of plant biology and Pérez-Paredes who worked on a corpus of English news language. Some of these studies are genre-based like Tribble's investigation (2000) into the lexical patterning of project proposals or Henry and Roseberry when looking at introductions to guest speakers and letters of application; others aim at a whole register like Nelson in his study on Business English (2001:93 ff.).

Concerning Nelson, his aim is two-fold; he does not only want to give a full account of Business English but also wants to find out how useful the information is which can be found in common Business English teaching materials. For this reason, he compares a corpus of commercially available Business English books with a corpus of authentic business materials and is able to illustrate the discrepancies between the two. This leads him to posit that most Business English materials are only suitable to a certain extent and that corpus investigations can offer a fuller picture. Bondi (2001:135 ff.) comes to a similar conclusion having examined how abstracts are dealt with in common EAP teaching materials and in a specific corpus of abstracts from different disciplines.
Besides the use of corpora for course and materials design, corpora in classroom use is a field of growing importance. Aston, as a scholar with a rich experience in bringing corpus data to the classroom of prospective translators, comments on this development as a phenomenon that is there to "enrich the pedagogic environment" (1997:261). Corpus data can be used to create exercises as well as being immediate subjects for students' research. One important work in the German context in this regard is Mukherjee (2002).

Furthermore, learner corpora continue to be the source of typical mistakes of a group of learners and can document over- and underuse of certain linguistic features. A recent small-scale example is Lynne Flowerdew investigating "lexical patterning, pragmatic appropriacy and discourse features" (2001:363).

And last but not least, parallel concordances have also continued to be a part of the small corpus applications searching for translational equivalents thus being mainly used in the training of translators (cf. Lawson 2001:279 ff. and Ghadessy/Gao 2001:335 ff.).

What all the above-mentioned approaches have in common is that ESP teachers started from a position where there was either no teaching and learning material for a certain domain-specific language or where they could not rely on the existing material. Using corpora to remedy this problem turned them into linguistic researchers and materials designers. Moshen and Ghadessy comment very enthusiastically on the fact that "the emergence of a new approach of language studies based on collections, or corpora, of texts, has revolutionised the teaching and learning of languages" (2001:XVII). This certainly holds true for the numerous investigations undertaken of which some were mentioned above. But despite the fact that the teaching and learning needs of increasingly diversified specific languages is ever growing, it has not yet become a mass phenomenon. Every ESP domain calls for a thorough knowledge basis for teaching and learning in order to meet their specific needs. Tribble (1997) said "that the most useful corpus for learners of English is the one that offers a collection of expert performances in genres which have relevance to the needs and interest of learners". And these "expert performances" were focused on in the design of ENGICOR with the hope that corpus data will "eventually lead to better informed choices in vocabulary selection, particularly in the staging and sequencing processes" (Pérez-Paredes 2003) for environmental engineering English. The following paragraphs will therefore first report on a needs analysis of the target situation of the engineering students before it goes on to describe the rationale behind the making of ENGICOR.

3.2 Needs Analysis

3.2.1 Preliminary Considerations

Having in mind the aim to specify language-learning objectives with the aid of corpus analysis, I now want to tackle the issue of how to go about preparing the selection of corpus material. Which material should be included? Which parameters determine the choice of material? These questions need to be answered before starting the corpus compilation.

Bowker comments on the importance of the selection process as follows: "Needless to say, that the value of what comes out of a corpus is heavily dependent on the quality of what goes into a corpus" (2000:22).

First of all, the corpus must be suitable for deducing information on lexical patterning and word-formation, as it is to serve as a basis for informed decisions in curriculum design of a specific language course, i.e. defining its learning goals and helping to decide on the methods and materials used. For this reason I need to look at the future professional situations in which the students will use their language and which topics they need to be able to understand, write and talk about, and subsequently which lexical and morphological patterns are used in the language. As Flowerdew puts it "the starting point is a corpus of written and/or spoken texts from the target communicative situation" (2001:71).

"A great deal of language that people are exposed to every day is very routine and predictable, just as are the situations they encounter." (Nattinger/DiCarrico 1992:27). The suitable corpus needs to bring these patterns to light and illustrate their use to allow generalisations for the specific language of environmental engineering English.

To achieve the above-defined aims, information on the target situation was needed. Needs analyses tha determine the target situation in which learners need to apply their language skills have become a common tool to prepare decisions in programme design (cf. Brindley 1989:63). In order to find out about the specifics of the target situation a questionnaire was designed and sent to 48 companies which had at some point employed a graduate or provided an internship for students from the course of study for which the English curriculum will be designed. On the one hand, the information hoped to be gained from the results of the questionnaire is the basis for the decision about which topics and text types to include in the corpus. On the other hand, the linguistic corpus investigation to follow can be considered as an extension of this target situation analysis. The picture will only be complete if situational and contextual information is complemented by linguistic information. They provide vital facts for the ESP-teacher.

3.2.2 The Questionnaire

Hutchinson and Waters state that a needs analysis has to find out "what the learner has to know in order to function effectively in the target situation (1987:55). This information can only be provided by people who have a profound knowledge of the situations in which students and graduates of environmental engineering will find themselves: experts in this field. The questionnaire is a pragmatic and practice-driven means to get information from these experts about the language situations and language activities in which they use the English language. Questions were asked with a view to ascertaining the following information:

- which field their company operates in
- whether staff members need English in everyday job situations
- in which departments the language is needed and to what extent
- which language skills are practised and their relative frequency
- which types of documents have to be dealt with
- which topics are predominant

- how they evaluate the language skills encountered

- what they think about the importance of mastering specialist language in comparison to general language.

Most of the questions used were closed questions in order to allow for quick processing of the questionnaire by the companies. But with most questions the opportunity to add comments were given but not widely used.

Of the 48 questionnaires sent out, 26 were returned which can be considered as a very good rate which might reflect the considerable importance of the question of foreign language skills in job situations.

Given the nature of the questions and the relatively small number of companies asked ("relatively" because the questionnaire was sent to all companies having ever employed a graduate or a student during the work experience semester from the course of study of Brownfield and Materials Recycling), the way to display the returned results can only be a descriptive one. Sophisticated statistics were not needed to establish patterns of useful information from the data.

3.2.3 Qualitative Analysis of the Questionnaire Data

19 out of the 26 companies which returned the completed questionnaire– according to their statements – clearly belong to the different branches of environmental engineering or at least cover one area that falls into this category. The branches mentioned were brownfield recycling, soil and groundwater protection, geology, hydrogeology, water and wastewater treatment, renewable energies, process engineering, metal and plastics recycling, waste management and waste treatment, and mine reclamation. A few companies gave just the umbrella terms plant construction or civil engineering from which sectors of environmental engineering can be assumed; others only said they were research enterprises without specifying the branch. Most can be described as companies and consulting agencies which provide services and know-how in the field of environmental engineering.

All of the contacted companies can be called small and medium-sized enterprises with the majority of them having 1 to 100 employees; just 3 had more than 100 staff. 23 of them had at that time employed students during their work experience semester; just five took on graduates after their studies (in 2003 when the first year group of the new course of study Brownfield and Materials Recycling graduated).

English was considered the most important and mostly used language in all companies, followed by Russian, French, Polish, Spanish, Czech and Italian in that order.

Of the 26 companies, 12 claimed to need English every day, 4 once a week, 2 once a month and 8 sporadically.

Department-wise, staff working at management level, on top of good general language skills, need environmental engineering English very often, followed by people working in projects, in after-sales scrvices and in marketing and sales. What was obvious from the questionnaires was that graduates are expected to be able to use not only technical language but also general and business English. Business English was considered most important for assistants, general technical English as vital for sales and after sales staff as well as in design. In particular, staff working in projects were considered to need mixed language skills.

Concerning the different language skills, reading was ranked as the most important followed by listening skills, then the ability to speak and writing skills were last.

In more detail, the ability to read and understand specialist texts, technical documentation and letters and e-mails with specialist content were considered most important. This was followed by understanding and reacting in telephone conversations with specialist content and the ability to write e-mails about a specialist subject. Less importance was assigned to the ability to write one's own texts with specialist content and reports, followed by translation of specialist texts. The ability to hold one's own presentations as well as the ability to follow specialist talks were of little importance but then these are activities which do not occur very often in their own right.

When looking at the answers given about the appropriateness of language skills of the current staff members, it was found that the density of information was not satisfactory to draw sensible conclusions. I only want to mention that the specialist language skills of environmental engineers (I also asked for managers, assistants, technicians, researchers, etc.) were considered satisfactory five times and not satisfactory six times. As this kind of question may have been seen as invasive by many people wishing to preserve the company prestige, some opted not to answer.

One batch of questions asked for the opinion of the person as to whether they agree or disagree with certain statements. The majority disagrees with the idea that a sound general knowledge is sufficient and that specialist language knowledge can be acquired during the work process; they make it quite clear that the mastering of general and specialist language is paramount. The majority claimed as well that in-service training is needed but most again specified that this is left to the staff members themselves to decide the best means of acquiring it. The opportunity of company courses was rare. Most companies consider the ability to act and react in the English language as an important factor for being competitive and expect the need for languages to grow in the future.

All of the above-mentioned facts are vital to know for the course designer who wants to create tailor-made courses for students who prepare for a working life in the respective branches of engineering. In order to create a corpus of specialist texts, it was furthermore vital to get to know which topics the experts from the companies have to deal with. The following fields of emphasis, which are very much in line with the earlier-mentioned branches of environmental engineering were mentioned:

- water and wastewater treatment
- brownfield recycling
- soil reclamation
- hydrology
- treatment of mining waste (revegetation and recultivation of mining waste dumps)
- reduction of emissions (air)
- materials recycling (electronics scrap, plastics, metals, debris)
- waste treatment (organic waste, biosolids)
- process engineering
- chemical analytics, soil sampling, groundwater sampling.

Summing up, it can be stated that there is a definite need for using the English language in work situations. Staff should be able to master everyday situations as well as to cope with specialist language. Readings skills were considered most important, in order to be able to understand specialist written texts closely followed by being able to act in specialised conversation.

3.3 Corpus Planning

Starting with these facts in mind, the question of what the corpus should look like came to the fore. The idea of collecting and transcribing oral texts was given up early due to the huge amount of time a transcription takes and the difficulty of getting hold of authentic spoken material. Consequently, it was decided that the corpus would include written material only. The next decision concerned the way of obtaining the texts. The most straightforward way for getting hold of texts in electronic format is the internet, so the web was chosen to serve as the main contributor to the corpus.

A range of search terms were collected on the basis of the topics mentioned in the filled-in questionnaires. Information from brochures as well as suggestions from lecturers from Brownfield and Materials Recycling completed the selection of search terms which were used to locate suitable texts with the help of the Google search engine. It was soon found that many of the texts were hosted at one particular website, that of the United States' Environmental Protection Agency (EPA). This website provides a vast array of information from environmental engineers to environmental engineers as well as to the public. As the information on this website was found to be very suitable for the purpose of creating a small specialist corpus and moreover guarantees a high quality of the information provided by a great number of experts, I decided to make intensive use of texts from this website. A few topics though could not be located on the EPA-website (i.e. mining reclamation or waste management). In such cases, the links collection of the EPA website were used as a starting point to look for related agencies or companies that deal with these problems. This is how a few texts of Canadian, English, German and Australian origin came into the corpus of otherwise mainly American-English texts.

3.4 Corpus Design

The foremost aim of this study is to establish a basis of lexical information of the most relevant language items of environmental engineering English to serve as a valid source for syllabus design of specialist English courses at University level. So one step to achieve this aim is to look for suitable specialist texts in which the domain-specific language use is present. Consequently, the information that results from the questionnaire sent to the companies must be converted into action.

However, from the feedback of the questionnaire it was clear that getting hold of samples of every mentioned text type was an impossible task given the time restrictions and availability of the material.

Taking the notion of topic as a starting point is an approach criticised by Sinclair, who holds that "it is clearly an internal matter, because the topic of the text is defined by the way in which language is chosen and used in the text;... The problem is twofold. One is of circularity, mentioned above. ... The other problem is that there are as many classifications as there are researchers" (2003:172). Sinclair warns that corpus designers should be aware of subjectivity. He furthermore includes another useful idea for the corpus compiler: the idea that texts should be chosen because of the "social role" they play within the context of communication.

With these notions in mind, the quest for texts to be included in the corpus started at the website of the Environment Protection Agency, a US government body that provides information sheets, technology fact sheets and reports of various kinds to the environmental engineer but also to the interested American citizen, a social role which goes hand in hand with the future communicative situations in which the students have to be able to participate. The topical keywords from the survey were located on this website and suitable texts were extracted for inclusion in the corpus. Where the information on this website did not suffice, the few hints from the website and the topical keywords from the survey served as useful pointers as to where to start Google searches for more material.

On top of topic relevance, other selection criteria which were considered in the process of corpus design were authenticity, the date of publication and a wide variety of text types. Authenticity implies that the texts were written by an acknowledged subject-field expert in the field of environmental engineering. It can not be ruled out that some texts were not written by native speakers, as the internet does not provide information on the authors. But as most texts were extracted from the website of a recognised professional organisation, i.e. the EPA, in general, it can be assumed that the level of language on such a website would be suitable for the purposes of my research. As the EPA website was not sufficient to cover all topics, a few English texts of German origin were included as well but it was felt that the quality of translation was very high.

With a view to achieving the aim of describing the most up-to-date situation of environmental engineering English, only texts from the 1990s up to 2003 found inclusion in the corpus.

The text types encountered on the EPA website were foremost reports, conference papers and technology fact sheets, mostly written by subject-field experts aimed to be read by subject-field experts. A few reports were written by students of environmental engineering and the corpus also includes some citizens' guides; fact sheets written by experts to be read by laymen.

What was clear from the questionnaires is that students and graduates from such an innovative field of engineering will be expected to be able to read and understand English specialist texts like the above-mentioned reports and fact sheets.

Another source of information are web journals which were also included in the corpus. And last but not least, 8 chapters of an introductory textbook on environmental engineering were added as well.

The texts from the internet were downloaded, stripped of the HTML-format and stored in ASCII-format. Pagination, tables, indexes and figures were eliminated. The textbook texts were scanned, also stored in text-format and proofread and compared with the original. In the end the number of texts amounted to 440 of various length; only full texts were included.

In the end, the corpus comprised 2 million running words which were presented in 6 subcorpora. These subcorpora on the one hand represent the main topics the students of environmental engineering have to deal with at our university: Mining Reclamation, Soil Reclamation and Waste Disposal and Recycling. On the other hand, the subcorpora reflect various text types, like the subcorpus of web-based journals, that of technology fact sheets and that consisting of an introductory textbook. Within the text-type based subcorpora, efforts were made to ensure that at the same time the texts reflect the above-mentioned topics.

All in all, the subcorpora comprise one section of mining reclamation material (approx. 200,000 running words), one of soil remediation material (approx. 300,000 running words), approx. 200,000 words of technology fact sheets, 120,000 words from an introductory environmental engineering textbook, 400,000 words on waste treatment and recycling and approx. half a million words from web-based journals, all together a corpus of 2 million running words.

The following quotation by Atkins et al. contains a very important statement for the small corpus designer: “In our ten years of analysing corpora for lexicographical purposes, we found that any corpus – however unbalanced – to be a source of information and indeed inspiration” (1992:6) and they go on to say that “knowing that your corpus is unbalanced is what counts” (1992:6). This opinion certainly does not disregard the task of careful planning and devising the corpus. The way ENGICOR was designed consisting of professional language used by people belonging to the discourse community of environmental engineers, ensures to provide information about the most important words and their lexical patterning as well as morphological information.

3.5 The Field of Environmental Engineering

The branch of environmental engineering evolved from Civil Engineering. It was not called by this name before the 1960s although some "early environmental engineers" left evidence of their work when building for instance the

Roman aqueducts. Later, poor sanitation causing epidemics in urban areas and the combustion of coal during the Industrial Revolution in England triggered the first reflections on environmental problems. Today environmental engineering comprises more than solutions to water and wastewater problems. As all human activities result in some sort of impact on the environment, people working in the field of environmental engineering also have to deal with air pollution, waste management, recycling techniques and brownfields and mining reclamation, to name but a few areas. Environmental engineering focuses on the development and application of technologies which aim to minimize the negative effects caused by the ever-growing population on earth. Environmental engineers design and operate treatment processes and monitor environmental quality and the availability of natural resources.

Environmental engineering is based on the fundamental principles of biology, chemistry, physics, and mathematics from where it borrows lots of its vocabulary. Further fields that contribute to the wealth of knowledge of environmental engineering are public health, ecology and meteorology. One of its youngest branches, waste treatment and waste recycling made a further contribution to its knowledge base: mechanical engineering and process engineering began to play a bigger role.

The fact that environmental engineering is a fast-developing science receives confirmation from specialists themselves. In their text book on environmental engineering Weiner and Matthews admit that "any text in environmental engineering is somewhat dated by the time of publication because the field is moving and changing rapidly." (Weiner/Matthews 2003: preface)

3.5.1 The Domain-Specific Language of Environmental Engineering English

People working in the field of environmental engineering communicate with each other in every day professional conversation as well as in scientific written discourse. The language they use is a subset of General English in that it builds on all the features of the general language but where it differs from it is in the extent to which the different features are used. The most striking difference is on the lexical level. Environmental engineering English has its own specific vocabulary originating from morphological processes as they occur also in General English. Again there are morphological processes which are used to a greater extent in a specialist language (compare Chapter 4).

The function of every communication is the provision and exchange of information (cf. Sinclair 1996b), and so in environmental engineering English the work-specific situations prepare the ground for the use of language by subject field experts to talk about specific topics. It is these topics which are for many linguists the main denominator that defines a special language (cf. Cabré 1998, Williams 1999, Gledhill 2000).

Bowker and Pearson (2002:39) therefore define a specific language as "the language to discuss specialized fields of knowledge". Consequently, environmental engineering English is the language to discuss matters concerned with remediation and recycling technologies. What specific languages have in common is that in contrast to General English they make use of specific linguistic means (i.e. the central role of passive constructions in technical English), generate their own text types (i.e. the plea in legal English) and bring about their own patterns of interaction (i.e. expert conferences). Exchanging information by the help of a specific language aims at a precise and economic way of communication (cf. Fluck 1997) as the success of the engineering activities and processes worked on together largely depends on shared specific knowledge expressed in language.

When talking about specific languages, the concept of a term is often discussed in the same context. Terms are often said to designate "a concept in the specialized knowledge of a domain" (Meyer/Mackintosh 1996:259). For the present study, terms do not play any role, as it is not from a cognitive point of view that the domain-specific language is looked at but examined on a lexical level. Based on the extensive analysis of the 500 keywords, which include also words which could be called terms as well (i.e. *alkalinity, feedstock, leachate, ozonation*, etc.), it was found that they do not display a collocational or morphological behaviour that is different in any regard from words without terminological character (i.e. *air, compounds, location, odor,* etc.). Their usage profiles did not demonstrate any differences in the frequency or kind of collocational and word-formation patterns. Moreover, the usage profiles indicated that prospective terms also acquire "additional meaning as a result of being used in texts" (Tognini-Bonelli 2002b:79); that a technical term "more often than not, becomes a part of a phrase with a specific semantic preference and a specific semantic prosody" (2002b:66) as Tognini-Bonelli claims in her ground-breaking article on the context-dependency of terms in the language of economics.

The study of the 500 keywords, comprising so-called terms and non-terms, is further evidence for this line of reasoning and shows that it is as well in the immediate context of language, in its use, that terms like any other words acquire further components of their meaning. The meaning of terms as well as those of words have to be learned in contexts, a process that subject-field specialists go through when acquiring domain knowledge and which a language learner has to replicate in the foreign language.

In this regard it is important to mention that domain-specific languages are also sometimes called restricted languages (cf. Firth in Palmer 1968:98 ff.) in that their vocabulary and their other specific linguistic features are more limited than those of General English (cf. Teubert 1998). This is due to the fact that a restricted or specific language deals with a limited amount of topics. This phenomenon is helpful for learners and also one of the principles which make the description of specific languages a feasible endeavour. Compared to the 1 million word LOB general language corpus, a subset of ENGICOR with the same size displays only 33.7% of the total of different word forms which shows that environmental engineering English is restricted to a certain amount of words.

4 Chapter 3 - Analyses Planning and Selection

4.1 The Selection Process

"A language is so complex that selection from it is always one of the first and most difficult problems of anyone who wishes to teach it systematically" (West 1953:V).

This quotation from Michael West reflects the problematic nature of the process of analyses planning. The first question to answer was which words are the ones which according to Firth are the most central, the most pivotal in the language of environmental engineering English.

The corpus features 33,105 types (single word-forms) of which 10,749 are hapax legomena (words which occur only once in the corpus). As not all words are equally important, a selection from the whole corpus has to be made. About 30 occurrences of a word-form are said to be needed to make a viable analysis of it with the help of a concordance; 5,473 word-forms occur at least 30 times, which still goes beyond the capacity of a single researcher. The frequency list (made with the WordSmith 4 software) displays the common picture: function words are most frequent and appear in the most prominent positions (the first ten most frequent words are: *the, of, and, to, in, a, is, for, that, be*). No effort needs to go into working on these words, as firstly their use should be known from school, and secondly they only provide support for the content words in the specialist texts. Nevertheless, ENGICOR starts showing its profile in its wordlist when for instance *water* is the first content word to appear in position 22, *waste* in 28, *system* in 31 and *site* in 42.

4.1.1 Word Lists

However, starting from the frequency list of ENGICOR alone was not considered suitable. So, the next thoughts addressed the question of how to get rid of the function words. As the idea of how to achieve most in language learning in proportion to the time and effort students spend on is not new; there have been numerous attempts to establish wordlists of frequent words which cover as much text as possible. One of the oldest studies in this field is West's General Service List (1953). The 2,000 words of the General Service List are said to cover around 80 % of the running words in academic texts (cf. Chung / Nation 2003), as they are the ones that work the hardest in any kind of texts. 80 % text coverage is not enough to ensure a comprehensive understanding of a text. In her study, Laufer found that "(a) chance to become a 'reader' is significantly higher if the lexical coverage of the text is 95 % and higher" (1988: 319).

In order to cover more ground than the 80 %, the Academic Word List was created by Coxhead (2000). This list comprises 570 word families of words that were extracted from a wide range of academic texts. The criteria of frequency and range of these words in a number of academic sub-disciplines were used to extract these words.

The idea was to deduct the words of the General Service List and the Academic Word List from the ENGICOR frequency list which would leave the filtrate consisting of environmental engineering English words plus a number of very rare words and proper names, etc. that had to be hand-edited. When going through this procedure, it was found that valuable words like *mine* as a homonym, and *water* and *waste* were eliminated from the resulting list, which counteracted the intention of keeping the most important words.
One could argue that these admittedly frequent and well-known words should be known to the students anyway but at this point I again want to draw my readers' attention to the principle that language mainly consists of larger chunks of language, the multiword items. The aim of establishing a word list is only a supportive step in this process, as the most frequent words are also considered to form the most frequent multiword items.

4.1.2 Keywords

For this reason, another operational procedure had to be found, which was to employ the keyword function of WordSmith 4. When Firth (in Palmer 1968:106) spoke of keywords he certainly was not anticipating of computer-extracted ones. WordSmith 4 calculates keyword lists by comparing the frequency of word-forms from one corpus with the frequency of word-forms from another. The resulting list shows the words that occur with a significantly higher frequency in the environmental engineering English corpus.

The British National Corpus (BNC), a corpus of 100 million running words of General English was chosen to serve as a reference corpus.

The result of the keyword comparison between ENGICOR and the BNC was satisfactory for my aims. Looking only at the first 20 positive keywords that occurred significantly more frequently in ENGICOR than in the BNC (*waste, water, tailings, wastewater, site, treatment, soil, landfill, groundwater, material, contaminant, concentrations, percent, flow, contaminants, effluent, recycling, chlorine, system, solids*), these words can be said to genuinely "smell" of environmental engineering English. The full keyword list can be found in Appendix A.

For this study, I decided to analyse the first 500 of these keywords in detail (for an example of such an analysis see 4.2).

When looking at these words, it is obvious to the reader that they are a mixture of technical and subtechnical vocabulary, in some cases even function words are left in. Still, the present approach is a step along the way to achieving better results in the text coverage by establishing the most suitable word-forms and consequently the collocational relationships these word-forms build.

From the point of view of this study it is not considered as a drawback that the list consists of technical words and subtechnical words alike (for differentiation of these terms see Curado 2002). It is felt that distinguishing between technical and subtechnical words is not a sensible thing to do in the light of teaching the subject-specific language to students. Anyway, decisions on which words are more the centre of interest are based on the relationships with other words with which they will function more closely.

Therefore the division between technical and subtechnical words did not seem appropriate as a multiword item can consist of both at the same time (i.e. *remove contaminants/impurities, (heavy) metals/nitrogen/nutrients/particles/solids/ VOCs, or contaminant/inhalation / radiation/ UV exposure*, with *remove* and *exposure* being listed in the AWL and their collocates being clearly words specific to environmental engineering English). The few function words that the list displayed were partly left as well, as they are for instance an indication of syntactic particularities of scientific and technological language in general.

The keyword list was manually edited to remove items which did not lend themselves to fruitful examination like *et, al., #*, country-specific abbreviations like *EPA* and *USEPA*, a few measuring units like *tons* and *gallons*, and auxiliary verbs like *can, is*, and *are* (high up in the keyword list pointing to the fact that the use of the passive voice is predominant in the investigated texts).

For the sake of this study, the first 500 word-forms of the keyword comparison between ENGICOR and the BNC were chosen with a view to making them the subject of a detailed analysis on the lexical (collocation), lexico-grammatical (colligation) and morphological level and in certain cases also to investigate the semantic and pragmatic level (semantic preference and semantic prosody) (cf. Sinclair 1996a). As this is the main ground to be covered by the further analysis in the course of this study, the question that remains to be answered is to what extent the choice of the first 500 words of the keyword comparison is justified in terms of learning effort and extent of the text coverage of these 500 words. How reasonable is it to concentrate on words from a keyword comparison between a specialist language corpus and a general language corpus?

By the way, 500 words is an arbitrary number. 500 lexical items were deemed to be feasible to make analyses within the scope of this thesis.

4.1.3 Text Coverage

In order to find out about the usefulness of the first 500 keywords in terms of their text coverage and subsequent balance of learning efforts and learning outcome, the Vocab Profile program (Cobb: Web VP v 2.0) was used to determine the text coverage of words from the GSL and the AWL in ENGICOR. As this program can only compute 28 files at a time, a selection from various subcorpora was chosen to serve as a sample of the whole corpus. The results are displayed in the following table:

Tab. 2: Results from Vocab Profile

text	**GSL 1-1000 in %**	**GSL 1001-2000**	**AWL**	**off-list words**
mining j	64.35	7.92%	9.34%	18.38%
mining n	67.04	5.78%	8.96%	18.22%
mining ww	66.81	6.47%	9.27%	17.46%
cigui 22	73.69	9.51%	7.09%	9.70%
cigui 10	69.81	7.39%	7.64%	15.16%
cigui 8	71.66	7.96%	7.15%	13.23%
student papers b	65.68	6.20%	11.09%	17.03%
student papers f	61.86	6.55%	9.06%	22.52%
student papers k	60.30	3.46%	10.34%	25.90%
remediation 30	64.32	4.00%	12.51%	19.17%
remediation 36	70.47	5.10%	11.28%	13.15%
tfs biosolids d	66.36	5.47%	12.32%	15.85%
tfs combined d	63.76	9.00%	10.14%	17.09%
tfs decentral i	64.54	8.69%	9.91%	16.86%
tfs waste water k	66.96	7.64%	9.01%	16.39%
thesis b	61.46	4.77%	12.59&	21.17%
textbook chapter 2	72.09	5.62%	8.65%	13.64%
textbook chapter 8	62.90	8.07%	13.39%	15.64%
chelsea 11	66.82	6.95%	13.08%	13.14%
chelsea 23	72.74	6.96%	7.44%	12.86%
chelsea 42	67.50	6.60%	9.72%	16.17%
green dot packag.	68.82	4.79%	11.43%	14.96%
waste glass densif.	67.15	8.48%	10.53%	13.84%
waste organics co.	69.03	5.29%	10.87%	14.81%
erosionMaiJun012002	68.51	4.89%	9.29%	17.41%
MSWMarApr022000	67.42	4.40%	14.07%	14.11%
polleng2000 02-1	63.43	4.77%	18.02%	13.78%
solidwasteAprMar2000	67.66	6.54%	9.02%	16.78%

average	**66.89**	**6.4025**	**10.4717**	**16.2292**
GSL gesamt		**73.2925**		

It was found that words from the GSL account for 73.3% of the text coverage which is less then the around 80% Nation established for a variety of academic texts (Nation 2001). The AWL words account for 10.5% which is very much in line with Coxhead's claim (2000:213) that her AWL covers around 10% of academic texts. This leaves a percentage of 16.2% of words in ENGICOR which must consist of environmental engineering English-specific words and a so far undefined rest. For illustration, a few of these off-list words are shown below. Most of them clearly belong to environmental engineering English.

Tab. 3: off-list words

• beneficiation	• ores	• profile
• EPA	• lode	• placer
• zinc	• update	• conservation
• congress	• hazardous	• exemption
• metallic	• asbestos	• overburden
• uranium	• unwarranted	• infeasible
• slag	• tailings	• timing
• technologies	• NPL	• thermal
• conductivity	• alloy	• utensils

To find out about the consistency of the remaining 16.2%, the keyword list of the first 500 words and the information on their individual text coverage was used to answer this question. All 500 keywords – which are chosen to be the subject of this study – have a text coverage of 22%. In this number, words which occur also in the GSL and the AWL are still left in, and were consequently taken out in a next step. The 500 first keywords minus the GSL and AWL words showed a text coverage or - as Nation calls it: vocabulary load – of 10.5%.

So, adding all these findings, words from the GSL, the AWL and the remaining body of the first 500 keywords amount up to 94.3% text coverage, a percentage that gets very close to the 95% Laufer claims to be necessary to understand a text successfully (cf. Laufer 1988:319). So concentrating on the first 500 words from the keyword comparison of a specialist corpus with a corpus of General English seems to be a worthwhile endeavour for the course designer. In terms of concentrating on these words within a time span of five semesters also seems feasible enough.

4.1.4 Detailed Consistency Analysis

However, the 500 keywords chosen for analysis were analysed again with the help of the detailed consistency analysis in WordSmith 4. This was done to find out whether there are word-forms that predominantly occur in one subcorpus rather than in the others. In this analysis, one subcorpus at a time was compared to the whole ENGICOR corpus. It aimed at finding some categories based on topic or text type which could help in syllabus considerations when, for instance, the language of mining reclamation displays other multiword items than those related to soil reclamation in general and which word-forms and multiword items occur with a high frequency over a wide range of text.

The results of this analysis confirmed the hypothesis. Of the first 500 keywords 195 showed a clear preference for one or two of the sub-corpora. To illustrate this, only a few words that predominantly occur in the mining reclamation subcorpus need to be mentioned here (*tailings, ore, runoff, acid, leaching, impoundment, discharge, drainage, seepage, dam, ponds*) compared to the waste and recycling subcorpus (*recycling, recycled, compost, composting, reuse, HDPE, plastic, MSW, feedstock, fiber, scrap, packaging*).

At this stage, a weak point in the corpus design was detected. Although the subcorpus of web-based journals is the biggest of all sub-corpora, it yielded the smallest number of topic-related word-forms. That has to do with the fact that it first comprises journal articles from a wide range of subjects relating to many different topics in the corpus, but foremost with the fact that it consists of a large number of texts. Within so many texts, not many single word-forms can become prominent enough to be included in the topic-based detailed consistency analysis, an unexpected extra finding in this procedure.

In sum, 500 keywords are available at the present stage for further analysis of their major structural characteristics.

4.1.5 Representativity

In 4.1.3 the need to justify the choices made in corpus design has already been made obvious in the light of the claim that corpora should be representative to make valid statements about language. Representativity has become a much-debated concept not only in small corpus studies (cf. Williams 2002, Teubert 2003). It is felt to be an ideal which should be aimed at by comprising as wide a range of samples as possible within certain limits. A truly representative corpus of environmental engineering English would consist of every utterance ever made and every text ever written in this professional field, an unachievable task.

Atkins et al. comment on this dilemma: "in our ten years of analysing corpora for lexicographical purposes, we have found any corpus - however unbalanced – to be a source of information and indeed inspiration" (1992:6). What this quotation says about the work of lexicographers holds true for teachers as well. As no suitable teaching material exists to cater for the needs of specialist language courses there is a strong need for small corpora of these sublanguages. If representativity is not the foremost corpus design criterion, other criteria have to take its place. Other small corpus studies like Hänchen (2002) mention diversity of addressees, topics and text types; Curado (2002) brings topic relevance and updatedness, course syllabi and availability into play. In the case of ENGICOR all these ideas have been considered to a certain degree when choosing the websites for text extraction. However, when making claims on the basis of small corpus analyses the shortcomings of every corpus have to be kept in mind. And this is what this study aims to show in the field of environmental engineering English. On my way to understanding the lexical level of environmental engineering English, the question as to what degree ENGICOR is "representative" yielded the following food for thought: the subsequent paragraph will try to demonstrate to what degree working with small corpora is valid and justify it as a step in the right direction towards devising new ways in which specialist language can be learnt.

4.1.5.1 Corpus Comparison at Word Level

The cooperation with the companies did not end with the initial questionnaire. After the corpus compilation they were contacted again and asked to provide some written English texts that they came across during their daily work routine. Compared to the initial feedback where more than half of the companies filled in the questionnaire, the reply to this request was quite weak but a mini-corpus of eleven texts with altogether 45,600 words could be compiled. The topics ranged from brownfields and mining reclamation through biogas collection to river basin management, topic-wise clearly environmental engineering themes. The same is true for their text types which were mainly project work drafts, scientific reports, one call for project proposals and project work email.

This mini-corpus of texts provided by the companies was used to make comparisons with ENGICOR. First, the wordlist of the two corpora were compared with the help of detailed consistency in WordSmith 4. The investigation revealed that the mini-corpus comprised 195 words that did not occur in ENGICOR (text-specific abbreviations, proper names, genitives have been disregarded). Then the frequency of these words was determined in order to see how much text coverage these words would yield and it was found that this figure amounted to 0.85%. After having a closer look at the 195 words, they were divided into certain categories of which geographical names (Poland, Colombia), American English/British English variation, easy words (dear, umbrella, handout) and obvious language mistakes (shutted, criterions) were eliminated as they were not regarded to count as words that could have occurred in ENGICOR if it was better designed. This left the number of words in the mini-corpus which did not feature in ENGICOR at 95 and their frequencies revealed that they only account for 0.25% text coverage.

At this point it has to be mentioned that I am not suggesting that a student is supposed to know all word-forms occurring in ENGICOR (which amount to 33,105). Text coverage is meant here as a measure that does not include a reader but only the two corpora. Furthermore, starting the whole journey from the wordlists does not mean the unit of investigation is a single word but of course the multiword item and therefore words are only the foundations for the scaffolding by which the investigation is supported.

0.25% divergence is regarded to be insignificant by this study and shows that claims on corpus grounds about lexical patterning in environmental engineering English, for which the words are only a starting point, will be valid if judged by means of a type and token comparison between the mini-corpus and ENGICOR.

For lack of a better word, ENGICOR seems to be "representative" of a high percentage of environmental engineering English and as a small corpus it promises useful information about linguistic characteristics of this specialist language.

4.1.5.2 Corpus Comparison at the Level of Multiword Items

Now, digging one level deeper, the next question is whether ENGICOR and the mini-corpus display a similar correspondence when multiword items are compared as well. To this end, 15 of the most frequent content words in the mini-corpus frequency list were searched for collocations and multiword items and the usage profile format used to take a record of the results. Then the usage profiles of these 15 words (*water, waste, treatment, groundwater, environmental, area(s), plant(s), surface, quality, anaerobic, environment, management, urban, sewage* and *protection*) from both corpora were compared to each other, counting the instances of two-word combinations and multiword-combinations that featured in the mini-corpus but not in ENGICOR. The result was quite surprising: 18.7% of the two-word combinations from the mini-corpus were not found in ENGICOR and even 44.6% of the multiword items from the mini-corpus were not found in ENGICOR, which makes as a whole 35.9% of elements beyond the mere word level where the mini-corpus differs from ENGICOR. In order to interpret this major discrepancy and before casting serious doubt on ENGICOR, we have to take a closer look at the origins of the texts from the mini-corpus. As they were not written by native-speakers of English, the question is whether the difference originates from collocation errors made by the authors or whether the cause for their difference might lie elsewhere. For this reason, the concerned language items were searched for on the internet by the help of the search engine Google. All of them could be retrieved with a reasonable amount of hits, so that collocational errors as source for this difference could be eliminated. The decisive hint came from the domains of the websites where the language items could be located: most of them clearly belong to a language variety other than American English, some could be predominantly allocated to Ireland, some to Great Britain, some to Australia/New Zealand, others again more to non-English-speaking countries of European origin. These findings have clearly shown the limits of ENGICOR, claims of language patterns can only be made for American English.

A closer look at the language items in questions nevertheless is a bit more reconciling. As the single words of which the items are constructed also feature in ENGICOR, understanding their composed meaning is in most cases easy. To illustrate this claim *underground water bodies* from the mini-corpus corresponds to *groundwater plume* in ENGICOR, *groundwater status* to *groundwater quality* and *aerobic plant* to *aerobic digestion plant.*

4.1.5.3 Justifying the Work with Keywords

In order to find evidence for the suitability of the first 500 keywords, the mini-corpus of texts provided by the companies came into play again. In batches of 500 words, the first, second, third and forth 500 words from the keyword comparison were located in the frequency list made up of the texts from the mini-corpus. Figure 7 shows the first findings which reflect that more items from the first 500 keywords occur in the mini-corpus compared to the second, third and forth 500.

As a next step the frequencies of these items were established and their values can be seen in Figure 8, showing as well that the first 500 keywords also yield more tokens (how often the items occur) compared to the second, third and forth 500 keywords.

It is important to mention that with this rather crude analysis the aim was to try to detect a trend rather than absolute numbers. It would be interesting to see a more detailed analysis which would include lemmatisation and the elimination of function words. If the lexical items of both lists, keyword list and mini-corpus word list were lemmatised, a tidier picture could have been drawn as the students are supposed to understand words from the same word family. The inclusion of function words were seen as a disadvantage particularly when looking at

Fig. 7: type

Coverage

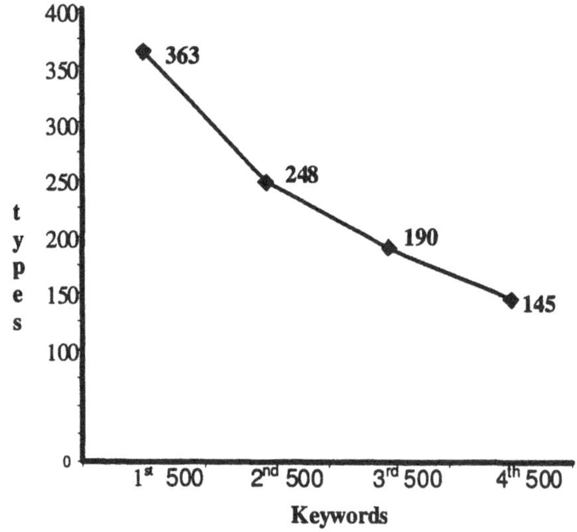

words like *the* which occurred 3611 times, and *which* occurred 1,406 or *of* which occurred 2,239 times in the frequency sums of the 500 keyword batches.

Fig. 8: token coverage

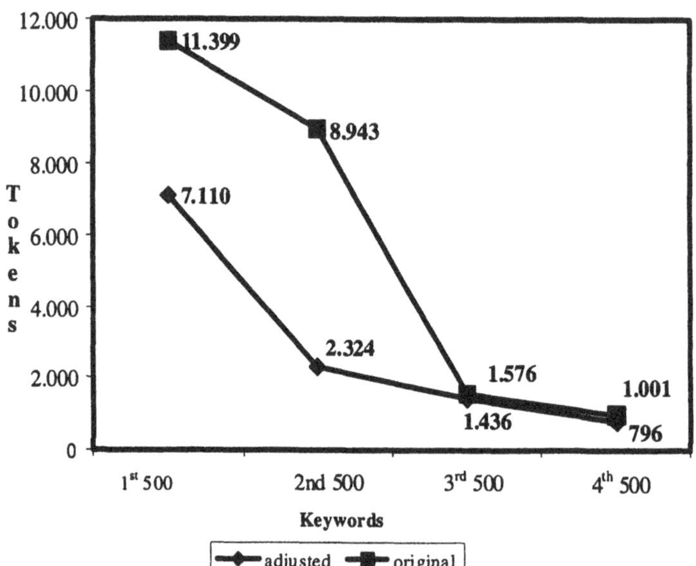

Therefore in the second graph of Figure 8 these very frequent items were eliminated which results in a slightly different graph which nevertheless follows the same trend.

The amount of times a lexical item occurs in a text, or in a corpus, as is the case here is called the text coverage of this item. In other words, spending time and effort on the first 500 words is more fruitful than on the second 500, which in turn is more fruitful than on the third 500 and so on. Figure 9 shows the percentage of text coverage that was achieved in the mini-corpus by the first, second, third and fourth 500 keywords.

Again, there is one version which uses the raw figures including also the most prominent function words and one version where these items were eliminated. The differences in the amount of text coverage by the first 500 keywords is significantly higher compared to the other batches of words. 15.48 % of text is covered by them, compared to 5.05% of the second batch, 3.12% of the third and 1.73% of the third (in the elimination version).

Fig. 9: text coverage

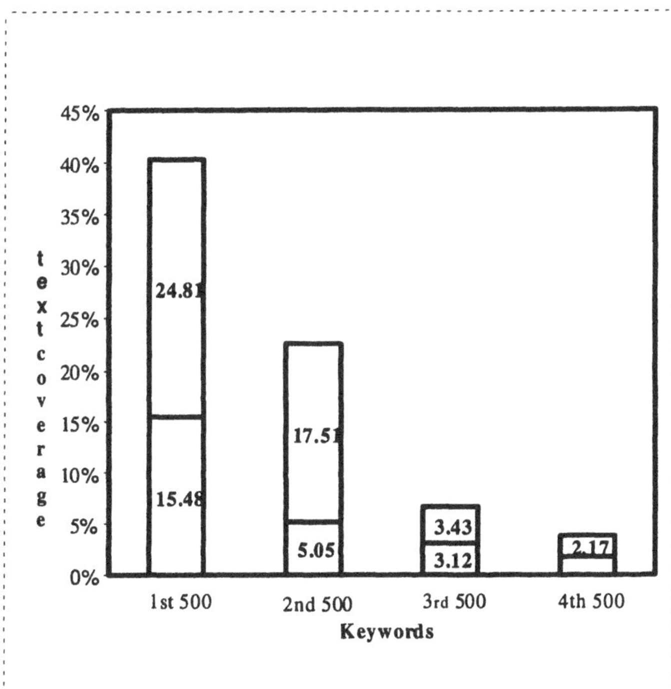

Looking back at the type-token ratio (a type being a lexical item in a corpus and the quantity of its tokens is how often this item occurs in the corpus) compared to the percentage of text coverage, it can be said that the quotient of text coverage to type frequency proves to be higher for the first 500 keywords and decreases along the line (being 0.04% for the first batch, 0.02% for the second, 0.016% for the third and 0.011% for the fourth. This can be considered as further evidence of the fact that concentrating learning efforts on the first keywords is time well spent.

4.2 Usage Profiles

The next step describes the search for a format in which a word-form can be presented as an extended unit of meaning. In other words a format where all lexical, semantic and morphosyntactic characteristics of the 500 keywords show can be recorded while being searched for by a concordancing program (WordSmith 4 in this case). It can thus be condensed to serve as an appropriate informant for later use in syllabus work. The aim is to find a form that portrays every word-form in accordance with the information on the use of language found in the corpus. This was particularly difficult as words from different word classes display different collocational behaviour but the format should be the same for all of them.

The characteristic features of use are expressed in Sinclair's model of an extended unit of meaning (cf. 1996a). His model shows what makes such multi-word items complete. Once all levels of this model are known by the learner, this lexical item should be available for the learner as a single choice and ready to be used in context.

There is first the level of lexical choices which features the preferred collocates; and secondly the level of syntactic choices which comprise the preferred colligations (lexico-grammatical patterns), both observable in concordances of the particular word-form. Furthermore, there is thirdly the level of semantic choices, the semantic preference, which reveals what semantic fields the item prefers; and last there is the level of pragmatic choices, the semantic prosody, which provide hints on how the item can be functionally interpreted. In short, collocation, colligation, semantic preference and semantic prosody are, according to this model, the four constituents of a lexical item which are functioning when this item is used in context and which should therefore also be considered when teaching and learning these items.

The following table which shows the schematic portrayal of such a lexical item serves to exemplify the form I chose to present the keywords:

Tab. 4: usage profile of *exposure*

exposure (488 x in corpus) noun rank 486 in keyword list (4)	
N-1 (1)	**N+1 (2)**
contaminant exposure(s) (8)	exposure assessment (7)
inhalation exposure (8)	exposure level(s) (5)
light exposure (6)	exposure pathways (13)
radiation exposure (5)	exposure period(s) (7)
sunlight exposure (6)	exposure route(s) (7)
UV exposure (13)	exposure time (14)
chemical exposure(s) (6)	
chronic exposure(s) (5)	
human exposure(s) (19)	
prolonged exposure (7)	
clusters (3)	
risk of exposure (6)	
route(s) of exposure (7)	
permissible exposure limits (6) (PELs)	
potential exposure pathways (6)	
long(-)term exposure (5)	
UV exposure time (10)	
1/2/3/5/10/.. minute(s) UV exposure time (12)	
exposure to (145) (air and water/(airborne) contaminants/benzene/carcinogens/ chemicals/dust/gamma radiation/high temperatures/humans/light/TCE/contaminated groundwater/microbes/the public/the weather/UV/...)	
corpus-attested word-formation patterns (5)	
derivational	**compounding**
exposable (2)	overexposure (1)
expose (15)	south-exposed slopes (2)
exposed (256)	
exposing (19)	
exposition (1)	
expositor (1)	
re-exposure (1)	
unexposed (6)	

In the format of this table the N(ode)-1 column (1) shows all words found to the immediate left of the search term *exposure*, and N(ode)+1 all words found to the right of the search term (2). In the cluster column underneath (3) all multiword items are collected and displayed. Frequency information (4) as vital corpus information is kept in this form as well. The number of occurrences in the corpus and its rank in the list of the 500 first keywords are to be found in the first line.

Collocational patterns are clearly to be found in columns 1 and 2 and as well in column 3. Column 3 also shows colligational patterns as in the example *exposure to*. In columns 1 and 2 all two word combinations containing mainly noun+noun combinations but also adjectives and verbs and to a very small extent also adverbs are collected. Column 3 is the place where all other kinds of syntactic patterns are collected. What is classified under cluster here are all multiword items (which means more than two words occur regularly together). This ranges from noun phrases (i.e. *catch basin sweepings, life cycle analysis*), verbal phrases (i.e. *remediate the site, minimize the effects of...*), adverbial phrases (i.e. *generated on-site, apply in situ*) to colligational patterns (i.e. *VOC from groundwater, exposure to UV, under acidic conditions*) and all kinds of prefabs involving for instance verbs, function words and adverbs as in *range from approximately A to B, BE commonly referred to* or *as discussed below*. For all two- and multiword items, their absolute frequency of co-occurrence is kept for reference behind the particular items.

Most two- and multiword items like nominal phrases or prepositional phrases have basic lexical meaning, others like *commonly, figure, table, appendix* and *below* are part of text stylistic and structuring elements. Whenever a word combination only concerns a particular word form of a word (i.e. just the singular or just the plural), the given examples will reflect this. Furthermore, in the cluster section, a record is taken of multiword items which allow for internal lexical variation of the pattern in question (i.e. *adsorption onto (aquifer) sediments/mineral surfaces/Fe oxides*, etc.).

For the example *exposure* (Tab. 4) , column 1 presents a tangible semantic preference for items from the semantic field of light and radiation (*light ~, radiation~, sunlight ~ and UV exposure)*. Semantic preference for contaminants, chemicals and other harmful influences can also be found in the cluster column (*exposure to benzene, carcinogens, dust ...*). If the category of semantic prosody is a valid one for application in a specialist language as well, then these examples also display a negative prosody. In detecting these features and making them available for the students in class, an extended unit of meaning is delimited and ready for use in speech and writing.

Corpus-attested word-formation patterns (5) are gathered as well at this stage of data collection and will be ready for analysis at a later stage of this study. The underlying idea here is that a relatively young and very dynamic engineering branch such as environmental engineering will produce many more specialist language items in the future and that this is reflected in the patterns found. Students have to be familiar with the most common word-formation patterns found in environmental engineering English in order to ensure understanding of these items. Students need to be aware that new items are constantly produced and that certain word-formation patterns yield a higher productivity than others.

Although most of the 500 first keywords are nouns, this table format is also used to capture the lexical, semantic and morphosyntactic characteristics of verbs, adjectives, adverbs and participles that occur to a lesser extent in the corpus.

What is archived in and shown by the usage profiles is not different from what is made obvious by a computer concordance. It is just that only the useful patterns are drawn from it and displayed in a rearranged format which makes future consulting of this information more accessible. The usage profiles though do not only profit from the advantages of computer concordances but also take on their drawbacks: language items which vary much in their distance from the search word might be easily overlooked and consequently not kept for record. Verbs in particular are likely candidates which are overlooked as I try to illustrate in the following concordance of organic contaminants. *Remove* in N-1 position occurs only 4 times which does not qualify it for inclusion in the cluster section of *organic* or *contaminant(s)*. A closer look at all forms of remov* no matter what their position reveals that *remove* is indeed an important collocate of *organic contaminants* which should be considered in class. This criticism of concordances which extends to the usage profiles as well concern all collocates of search terms whose variable distance to their nodes tend to result in them being overlooked. With verbs this seems to be a frequent case.

Fig. 10: concordance of organic contaminants

... Widely used to *remove* biodegradable [[organic contaminants]] and inorganic nutrients (e
... ic contaminants Treatment *Remove* other [[organic contaminants]] Treatment Polish organ
... Oxidation Used increasingly to *remove* [[organic contaminants]] from ground water and
... nd an activated carbon filter to *remove* [[organic contaminants]] and residual chlorine.
... processes are typically used to *remove* [[organic contaminants]]. These processes can inc
... of solvent extraction is for *removing* [[organic contaminants]] from aqueous solutions. T
... ... hing is mostly used to *remove* synthetic [[organic contaminants]], a type of contaminant
... technique for *removing* a suite of toxic [[organic contaminants]] from water. In air quality
... emove metals Treatment *Remove* volatile [[organic contaminants]] Treatment Remove ot
... or a typical P&T system where volatile [[organic contaminants]] *are removed* from groun

4.3 Analysis on the Level of Collocational Behaviour

During the process of putting the linguistic data of the 500 keywords derived from the computer concordances into the usage profiles, it was found that this form of analysis brought to light that the most frequent form of lexical co-occurrences are two-word collocations. Of these the majority were noun+noun combinations followed by adjective+noun combinations, which sounds natural, as the most common word class is that of nouns.

By going through the meticulous work of searching the corpus for the collocational profiles of 500 words and keeping a log of them in the usage profiles it was hoped that such a vast amount of data would reveal at least some underlying regularities which would start showing at least some kind of orderly behaviour for an otherwise so arbitrarily-behaving phenomenon as collocation.

4.3.1 The Subgroup of Chemical Nouns

For this purpose, the subgroup of the chemical nouns was chosen in order to see whether their usage profiles show similarities in their collocational behaviour and if so how representative they are. All 39 usage profiles of the nouns designating a chemical from *aluminum* to *zinc* were searched for common collocational patterns. The analysis showed the following:

1. In a node –1 position the following words were found:

Tab. 5: Chemical nouns

noun + chemical noun	
metal	oxide(s), sulfide(s)

adjective + chemical noun	
aqueous	ammonia, chlorine, ozone
chlorinated	ethylenes, hydrocarbons, phenolics
dissolved	Al, Fe, hydrogen, oxygen, ozone, Pb, Zn
gaseous	chlorine, ozone, sulfur
halogenated	aliphatics, hydrocarbons
inorganic	carbon, chloramines, halide(s)
liquid	chlorine, oxygen
molecular	chlorine, oxygen, ozone
organic	carbon, chloramines, halide(s), nitrogen
oxidized	iron, nitrogen, sulfur
residual	carbon, chloramines, chlorine, ozone
total	nitrogen, phosphorous, Zn

verb + chemical noun	
contain	chlorate, lead, oxygen, PVC, sulfides, sulfur
convert ~ to ~	ammonia to nitrate, nitric oxide to nitrogen hydroxide, TCE to ethene, calcium oxide to calcium hydroxide
extract	carbon, copper, lead, uranium
form	bromate, chloramines, chlorine dioxide, manganese dioxide
produce	chlorite, copper, ozone, pyrite
recover	copper, lead, uranium
remove ~ (from)	ammonia nitrate, chromium, copper, halogens, nitrate, nitrogen, perchlorate, pyrite

2. In a node+1 position the following words were found:

chemical noun + noun	
ammonia, chlorine, dioxide, methanol, nitrate, oxide, ozone, permanganate, peroxide, phosphorous	addition
carbon, molybenate, Cr(VI), Zn	adsorption
chlorine, dioxide, ozone	application
carbon, chlorine, hydrogen, nitrogen, sulfur, oxygen	atom(s)
carbon, chlorine, hydrocarbon, mercury, nitrogen, sulfur	compound(s)
copper, molybdenite, molybdenum, pyrite, sulfide, zinc	concentrate
aluminum, ammonia, bromide, chloride, chlorine, CO, copper, Cr, formaldehyde, hydrocarbon, iron, nickel, nitrate, nitrogen, oxygen, ozone, sulfate, TCE, zinc	concentration(s)
nitrate, oxygen, potash	consumption
aluminum, carbon, copper, nitrogen, oxygen, phosphorous, pyrite, sulfur, zinc	content(s)
chlorine, oxygen, ozone, pyrite	demand
chlorine, ozone, potassium permanganate	dose(s)
carbon dioxide, hydrocarbon, mercury, methane, monoxide, radon, sulfide	emission(s)
copper, uranium	extraction
copper, copper-molybdenum, pyrite	flotation
ammonia, chlorine, carbon dioxide, sulfur dioxide, hydrogen, methane, nitrogen, oxygen, ozone, radon, sulfide	gas
chlorine dioxide, methane, ozone	generation
ammonium, bromate, bromide, calcium, chlorate, chloride, chlorite, hydrogen, hydrochlorite, permanganate	ion(s)
ammonia, chromium, dioxin, hydrocarbon, iron, lead, manganese, nitrogen, oxygen, ozone, phosphorous	level(s)
nitrogen, phosphorous, potassium	loading(s)
copper, gold, molybdenum, oxide, phosphate, uranium, zinc	mine
copper, sulfide	mineral(s)
copper, gold, phosphate	mining
carbohydrate, chloramine, chlorine, hydrocarbon, hydrogen, oxygen, ozone	molecules
copper, oxide, silicate, sulfide, uranium	ore
ammonia, carbon, manganese, ozone, permanganate, pyrite, sulfide	oxidation
aluminum, calcium, copper, iron, manganese, nitrogen, sulfur, uranium	oxide(s)
carbon, iron, titania, zirconia	particle(s)
copper, Cr, hydroxide, iron, phosphate	precipitate
carbonate, hydroxide, sulfide	precipitation
copper, hydrocarbon, uranium	processing
copper, carbon dioxide, methane, uranium	production
copper, gold, methane, uranium	recovery
chlorite, chromate, methane, nitrate, sulfate, uranium	reduction
chlorite, lead, phosphorus	removal
ammonia, oxygen	requirement(s)

chloramine, chlorine, ozone	residuals
chlorine, sodium chlorite, copper, cyanide, hypochlorite, permanganate, phosphate	solution
carbonate, chloramine, chlorine, nitrogen, sulfide	species
carbon, chlorine	storage
dioxide, oxygen	supply
copper, sulfide, uranium	tailings
oxygen, ozone	transfer
carbon, chromium, oxygen, Pb, Zn	transport
ammonia, copper	value(s)
ammonia, mercury	vapors

Clusters	
ammonia, bromide, chlorine, hydroxide, methanol, pyrite	was added
pyrite, uranium	was recovered
ammonia, cyanide	was detected
chlorine(-)to(-)ammonia, C:N, carbon to nitrogen, peroxide:ozone, nitrogen to phosphorus	ratio
BE converted to	ammonia, carbonate, nitrate

What these tables show is that the class of chemical nouns behaves in a remarkably homogeneous way on a collocational level. Generalisations can be made up to a certain degree; i.e. it can be assumed that a lot of chemical nouns can co-occur with *concentration* or *molecules*, whereas other combinations are restricted by the laws of chemistry or other constraints of the real world (i.e. the number of substances to be found in the form of *tailings* is quite limited).

The category of semantic preference comes into play here again with the chemical nouns constituting the semantic set of possible collocates. There are always a certain number of chemical nouns which share a certain collocational frame. But as the example of *tailings* shows, the generalisation has its limit; and that even more so for the ESP-teacher than for the language learning students whose background knowledge of the specialist field might allow them to build more correct collocations. The above-shown attempt at the systemisation of local lexical and semantic information of chemical nouns reveals the above-listed tendencies but needs to be verified by a bigger corpus of texts belonging to domain-specific text-types of chemistry.

This can as well be continued on the word-formation level. The following table shows that chemical nouns also tend to co-occur with the same past and present participle as well as adjectives which together with an initial chemical noun form compound adjectives. This phenomenon is described more in detail in 5.2.4.1.3.

Tab. 6: word-formation patterns including chemical nouns

chlorine, hydrogen, petroleum, sulfur	-based
carbon, oxydydroxide, silica	-coated

cyanide, lead, mercury, perchlorate, petroleum, Zn	-contaminated
oxygen, pyrite, uranium	-depleted
cyanide, dioxine, zinc	-laden
copper, iron, oxygen, pyrite, sulfide, sulfur	-bearing
carbon, chlorine, lead, mercury, nitrogen, ozone, sulfur, uranium	-containing
hydrocarbon, PCB, TCE	-degrading
oxygen, ozone	-depleting
copper, lead	-leaching
copper, hydrocarbon, uranium	-processing
nitrate, sulfate	-reducing
CFC, lead, mercury	-free
chlorine, ozone	-resistant
copper, oxygen, ozone, silicate	-rich

These compound adjectives are not the only examples of chemical noun items which display similar morphological behaviour. Threads like *oxidation, ozonation, methanation, chlorination* and *chloramination* as well as *oxidizer* and *methanizer* can be named here as well.

Now the question is whether such shared collocational behaviour can also be found amongst the other usage profiles. Do certain lexical items have similar collocational profiles and if so to which extent? This will be the subject of analysis in the next paragraph.

4.3.2 Collocational Behaviour in General

To this end, 30 words were chosen and their collocational behaviour compared. The focus was on all collocations and multiword items like two-word combinations in the form of noun+noun, adjective+noun or verb+noun, all forms of three-word combinations (i.e. noun+noun+noun, adj+noun+noun, etc), four-word combinations, prepositional phrases, verbal phrases and so on. In that way the collocates of 10 academic English lexical items like *analysis, content, location, generation, method, area,* etc. and 10 technical terms like *biomass, aquifer, denitrification, leachate, oxidant, permeability*, etc., as well as 10 adjectives were listed to be compared.

As already stated for the chemical nouns, two-word collocations were the most frequent, more typically noun+noun collocations when the node was a noun, more typically adjective+noun collocations when the node was an adjective. It was found that there is no difference between the academic words and the technical words. Two-word collocations make up for the vast majority of the items, the rest can be said to vary a lot in terms that one item occurs rather frequently in a prepositional phrase whereas a specific verbal phrase is typical for another.

This confirms findings from other studies like Smadja who comments on the arbitrariness of collocational behaviour as follows: "Collocations vary tremendously in the number of words involved, in syntactic categories of the words, in the syntactic relations between the words, and in how rigidly the individual words are used together, ..." (1993:144).

This holds also true for words looked at from a more intuitional angle; lexical items which belong to a semantic set should have at least some similarities and thus provide some usage guidelines for language learners. Like the class of chemical nouns, it was thought for instance that the class of recyclable materials would show similar tendencies and their concordances and usage profiles were searched for hints of this. The outcome was meagre. Apart from very basic concepts like *ground wood/rubber/ glass, scrap rubber/glass* and *recycled glass/rubber* etc., no more common collocates were found (it has to be noted that the frequency of co-occurrence was set to 5, so other combinations might exist but were just not frequent enough to be noted). Where there was great correlation again is in the form or structure of collocations, i.e. the high tendency to nominalisation.

The infrequent cases where lexical items share the same combinatorial characteristics occur when they also share a part of the context of situation: like *density* and *gravity* both being able to collocate with *separation* as they designate the underlying technological principle. It can be said that *density* relates in the same way to *separation* as *gravity* does. A more detailed look at a similar example differentiates the view again: The group of words designating process parameters often collocate with *high* or *low* as in *high* or *low density, alkalinity, permeability, concentration* but others take more specific modifying adjectives as in *acute toxicity, excess moisture, full* or *limited capacity*. You can furthermore *increase* or *enhance strength*, but *improve* or *optimize performance, reduce* or *exceed capacity* (these assumptions again focus on the most frequent collocates and not on what is possible).

There are certainly studies which examine a corpus to find the different types of collocations as for instance verb + adverb, adjective + noun, noun + genitive object, etc. (cf. Siepmann:2004) or similar information found in collocational dictionaries (Hill/Lewis:1997, Benson et al.:1997). It was however felt that for the purpose of this study, which is geared towards teaching and learning of a specialist language, this kind of information would not yield any significant outcome which could be used in the teaching and learning context as it is not important to know which kinds of classes the collocations might occur in more frequently but to be able to analyse that collocation which is focused on in the immediate classroom context.

What this investigation has shown is that the environment of every word is truly unique and no generalisations can be made. It stresses the importance of learning and knowing the "company a word keeps", as similarities account only for a very small part.

4.4 Networks of meaning

In 4.1.4 detailed consistency analysis has been applied to find out which of the 500 keywords occur significantly more often in one subcorpus than in the others and which words occur more evenly throughout the whole corpus. As a result the subcorpus wordlist in Appendix A reflects the categorisation based on this detailed consistency analysis. This was done to help curriculum decisions when working for instance on a particular topic starting with the central not with the peripheral words and first of all to find out which words can be considered as central.

At this stage, I want to mention another method to approach the notion of topicalness or aboutness and ask how certain very central terms relate to each other in text. This can make intuition superfluous when deciding which language material to begin with when working within a special topic.

The idea underlying this work is that lexical items have relationships on a semantic level which are expressed by the fact that within a certain topic some are more likely to occur than others. This phenomenon has been looked at from several points of view, as for instance Halliday and Hasan investigated lexical cohesion and established "the cohesive effect achieved by the selection of vocabulary" (Halliday/Hasan 1976:284). Stubbs (2001) showed how collocations contribute to text cohesion and together with Barth he used "recurrent phrases as text-type discriminators" saying that "technical words may signal the topic reliably, but give only a hint about text-type" (2003:66). What these studies have in common is their seeking of a "global pattern of textual organisation" (Phillips 1985:4) and that the "choice of vocabulary in a text is largely a function of the subject matter" (Phillips 1985:31).

For the purpose of this study, the fact that cohesion gets created by vocabulary items has to take a backseat to the more important question of how such a lexical network can be detected and what it tells the curriculum planner. The tool to achieve this is again WordSmith 4.

In a first step the three subcorpora 'mining', 'remediation' and 'waste and recycling' were chosen as they are those subcorpora which reflect the notion of topic best compared to the textbook subcorpus or the subcorpus of 'webjournals' which consists of topic-wise more heterogeneous materials. The texts of the three chosen subcorpora were subdued to a key keyword analysis compared to the whole corpus and in this way a database of key keywords was established together with their associates, that is words which are key in the same texts as the key keywords (a key keyword being a word which is frequent over a number of texts, cf. Scott WordSmith 4, help files). The following tables show the results of this analysis. In table 7 one can see that the key keywords which are key in a majority of mining texts are *mine, tailings, mining, rock* and *acid.* The words listed below them are their associates.

Tab. 7: key keywords and their associates for the mining subcorpus

key keywords	**mine**	**tailings**	**mining**	**rock**	**acid**
associates	underground	layers	pit	waste	drainage
	piles	impoundment	mineral	seepage	minerals
	AMD	pit	mines	underground	piles
	pit	dam	minerals	pit	waste
	tailings	**mining**	**tailings**	piles	seepage
	acid	**rock**	**acid**	**tailings**	**mining**
	mining	**acid**	**rock**	**mining**	**tailings**
	rock	**mine**	**mine**	**acid**	**rock**

Tab. 8: key keywords and their associates for the remediation subcorpus

key keywords	**groundwater**	**chemicals**	**soil**	**polluted**	**cleanup**
associates	site	methods	up	pollution	contamination
	soil	pollution	**groundwater**	methods	methods
	injection	up	clean	**soil**	up
	contaminants	**soil**	**cleanup**	up	clean
	zone	**cleanup**	**polluted**	**cleanup**	harmful
	chemicals	clean	harmful	clean	**soil**
	contaminated	harmful	**chemicals**	harmful	**polluted**
	polluted	**polluted**		**chemicals**	**chemicals**

Tab. 9: key keywords and their associates for the waste and recycling subcorpus

key keywords	**product**	**material**	**recycled**	**plastic**	**process**
associates	consumer	applications	processing	thermoplastics	applications
	markets	melt	consumer	density	testing
	molding	content	molding	products	materials
	products	equipment	**plastic**	strength	paper
	plastic	molding	scrap	manufacturing	content
	recycled	materials	applications	mixed	**material**
	material	**process**	**product**	**recycled**	**product**
	process	product	market	**product**	end
	content		markets	plastics	

For the purposes aimed at, these tables already reflect very nicely but on a very global level which are the most central terms when starting lexical work with the students on the topics of either mining, remediation or waste and recycling. They can serve as a kind of index for the texts occurring within these topics.

Scott comments on this as follows: "...and other texts in which the same word is key are likely to share concerns with the texts (especially if the texts have more than one key word in common)" (Scott 2001:109).

Through this study the analysis of keywords and their associates is considered as a first, global level of aboutness of text. The second more detailed insight into the aboutness – this time of individual texts – can be provided by the keyword analysis of a particular text compared to the whole corpus. And as this has found mention already in 4.1.2 I just want to illustrate that phenomenon by showing how a specific keyword list relates to the global aboutness of for instance the topic of mining.

Tab. 10: keywords from single mining text

mine	salmon	coho
AMD	dumps	forever
(BC)	(Coal)	(Rosemary)
river	(Mount)	ABA
mining	cyanide	millions
acid	(Bulkley)	silver
(Telkwa)	minesite	(Fox)
(Tsolum)	ore	conservationists
(Brandt)	fishery	(Skeena)
rock	million	copper
tonnes	steelhead	**tailings**
generating	waterways	centuries
(Equity)	waste	**mineral**
gold	till	streams
fisheries	proposed	says
(Britannia)	rivers	abandoned
mines	there	laws
(Manalta)	(Father)	into
drainage	(Ferris)	blended

This list of keywords contain words belonging to proper names (those put in brackets), words which occur very infrequently in the corpus like *till* (as in *glacial till cover*) and - most importantly - words which occur very often in this particular text as they are needed to talk about the topic of acid mine drainage. With a bit of background knowledge these lexical items already tell the story of the text which presumably is about *abandoned mines* and *minesites* where in the past *copper*, *silver* and *gold* were mined and nowadays *waste rock* left in *tailings* and *waste dumps*, *generating AMD* that is *acid mine drainage*, constitute the main environmental risk to *rivers* and *streams* and *waterways* in general and consequently to the *fisheries* in the area. There is mention of some sorts of fish like *salmon, coho* and *steelhead.* It can be assumed that the text is also about the amount of *ore* extracted as we have the word *tonnes* here as well. *Cyanide* might be talked about as one form of *AMD* and the role of the *laws* mentioned as well. Admittedly, some of the words do not ring a bell like *blended* and others which are function words and do not help to constitute the idea of aboutness. But as for the networking of the rest, it allows for a kind of anticipation of the topic before dealing with the specialist text and are consequently suitable for serving as the central words which need to be focused on, before, whilst and after the reading activities. It goes without saying that I do not suggest concentrating on the words only but on their collocational environments. And although the words are not collocates to each other they relate to each other in form of an "intercollocation" which create aboutness of text (cf. Phillips 1985:144). There is no doubt that a look at the keyword list allows an insight into a network of meaning, which makes a topic more accessible for teachers and learners.

The words put in bold in the table are those which have already featured amongst the key keywords and their associates, so we can now sense the more global character of these compared to the more specific character of the keywords of an individual text. It shows furthermore the dynamics of aboutness or topicalness of texts: there is a high likelihood that words like *mine, mining, rock, acid* and *tailings* occur in texts dealing with the legacy of abandoned minesites but the individual topic of the specific texts adds more aspects to it. Networks created by interrelated items relating to each other on a semantic level and thus forming lexical sets are probabilistic, i.e. no elements are obligatory when dealing with a text of the topic in question.

5 Chapter 4 - Word-formation in Environmental Engineering English

5.1 Preliminary Remarks

This thesis operates between the following two aspects; prefabricated and well-established chunks of language on the one hand and newly created language items on the other. Whereas the previous chapter dealt with the different lexical and meaning aspects of collocations, the current chapter will investigate morphological aspects typical for environmental engineering English. I will move from the phrasal perspective that focused on the interplay of words in collocations to the word level and how the structure of word-formation elements looks. Both have one thing in common, namely the fact that meaning is created in the situational and language context, but while language learners deal better with collocations in their natural state of raw chunks, productive word-formation patterns should at least be the subject of reflection in the language learning processes in order to be able to keep up with language change and facilitate the understanding of newly encountered words. Again both, collocations and word-formation products, have in common that they gain access more easily to the minds of the language learners the more often they are encountered in language use. Frequency though is now no longer of central interest (except for when I will look at the issue of productivity) in this chapter, because current trends in language growth and language change manifest themselves first in rarer events.

The fact is that language change is a fundamental staple of human society. Advances in science and technology contribute to the fact that language has to designate new objects and concepts. Environmental engineering, as a relatively new and innovative field of knowledge expansion will be examined for word-formation patterns that reflect the trends of accomplished language change and processes in action.

Language change is difficult to observe and could, not until the advent of corpus linguistics, have been systematically described. Large monitor corpora have made systematic and comprehensive description possible. ENGICOR as a small specialized corpus can only be a miniature image of a short period in the development of environmental engineering English reflecting the language of the text types collected but I claim that tendencies in the process can nevertheless be established: These findings can guide language learners in their future contact with this specialist language. I will describe a method that will bring to light aspects of lexical change typical for environmental engineering English by illustrating and analysing word-formation patterns found in the corpus.

Marchand (1969:2) defines word-formation as "that branch of science of language which studies the patterns on which a language forms new lexical units". The products of such word-formation processes are understood by language users on the basis of an implicit knowledge of the constituents of these processes, an ability a language learner has to acquire throughout the learning process. To this end, the elements that are active in the processes and their results need to be looked at. The results of word-formation processes are complex words produced by putting elements together or shortening existing elements. In English, we have bound elements like affixes and bases; free bases like in the case of *ion* and bound bases like in the case of *sol-* in s*oluble* at our disposal to combine existing language items into new ones. Such processes are subsumed under the headings such as compounding, combining and derivation processes.

Other changes are brought along by processes like blending, clipping, or abbreviation where parts of the original language items are either only omitted or omitted and the remaining ones newly combined. Conversion describes processes of word class changes, semantic change subsumes processes where language items acquire new meanings in new contexts. Several studies back up the idea that the use of existing resources of words is more common than the creation of new words (cf. Sager 1997:28). Furthermore, the motivation to form new items is usually divided into two (Bauer 2001:24) or sometimes three categories (Plag 2003:59-60):

- the need to designate new concepts (also called labelling function)
- syntactic recategorisation and
- expression of attitude

(the latter being the lesser mentioned one). In the course of this study, the labelling function was found to be the prevailing one, assigning syntactic recategorisation more to the stylistic field.

The present study prefers an evolutionary model of language change where new language items come into being by creation, which is a very rare event. The combination of existing elements is the next step and by far the area where most change takes place, as it is the infinite power of combinatory possibilities that provide for the dynamics of such processes. The third element which is only partially reflected in the above-mentioned syntactic recategorisation is the tendency of the English language to condense linguistic information again by abbreviating, truncating and converting (in the sense of Hickey's univerbation 2003:94, cf. 5.2.9.1.1).

The actual instances that will illustrate to serve the patterns found in environmental engineering English are found somewhere along this cline. So the labelling function and language economy are considered the driving force for the word-formation process which is achieved by either creating new words, combining the existing word-formation elements or shortening (and recombining) them.

The following chapter will identify and categorise word-formation patterns in environmental engineering English, highlight areas of preferred ways of word-formation and active change in the light of its use for the learner of that specialist language.

5.2 Taxonomisation Attempts of Word-Formation Phenomena

As part of preparing my own way of categorising word-formation phenomena present in the corpus, three different approaches to taxonomies were looked at: Caso's (1980), Algeo's (1980) and Bauer's (1994). The aim they had in common could be characterised by the question "(w)hat kinds of word making are being used and with what consequences" (Algeo 1980:267); Algeo and Bauer in the field of General English and Caso in the field of Scientific English. Furthermore, there will be mention of a corpus-based sampling method and its findings relevant to word-formation patterns (Renouf 1993). Firstly, their method of sampling the subjects of their study – the individual language items – should be described very briefly and then I will go on to compare and discuss their findings and derive my own method based on these grounds.

5.2.1 Dictionary-Based Approaches

5.2.1.1 Arthur L. Caso: The Production of New Scientific Terms

In his study, Caso concentrated on the language of physics and earth sciences and chose 4,300 terms from the Chambers Dictionary of Science and Technology edited in 1971. By using another dictionary, Webster's Third New International Dictionary of the English Language, he established the etymologies of 3,718 of the chosen words and assigned them to the six categories of his word-formation process typology: Semantic Change, Morphophonemic Change, Functional Change, Composition, Neologizing and Borrowing, and Miscellaneous, of which each category was further subdivided into types and subtypes (1980:103). Interestingly, he found the category of Semantic Change to be the one that yielded most new words. All words that "alter meaning but not form or function class" (1980:103) were assigned to this category.

The great number of words that underwent semantic change in order to fulfil new designation functions in the specialist language was followed by an almost equal amount of words that were created by composition processes of "elements already existing in English" (1980:103); such as affixing, blending, compounding, combining and acronymy and initialising.

Neologizing and Borrowing was the category that was third in its contribution to new words in the scientific fields of physics and earth sciences, as Caso found cases of borrowing and loan translation from German, French and other modern languages and also attributed loan words from Latin and Greek to this category.

He also found that the categories of Morphophonemic Change and Functional Change contributed to a lesser extent to word-formation processes. In Morphophonemic Change all instances of clippings were collected and Functional Change comprised all instances of word class changes, like nouns into verbs, nouns from an adjective, etc.

Caso does not define how he sees productivity in his article but as he assigned the single words to the different categories and worked out the percentages of the categories, it can be assumed that he thought of the category with the highest amount of members as the most productive and that it "should reflect fashions of scientific word formation" (1980:103).

5.2.1.2 John Algeo: Where do all the new words come from?

Algeo also goes for the dictionary approach as the basis for his sampling method. But this time General English is the subject of investigation. His study is based on the Barnhart Dictionary of New English which aims at reflecting new entries into the English language of the decade from 1963-1972. About 1,000 of the 5,000 dictionary entries were chosen in order to establish the word-formation patterns on which these were produced.

Unlike Caso, Algeo had no explicit semantic category but includes semantics in the Shift category and concentrates more on the origin and form of the language items. He too arrives at six categories with types and subtypes: Loan words, Shortenings, Composites, Blends, Shifts and Unknown Etymologies. Methodology-wise, each of his chosen words has to undergo a process where the question of whether it is of foreign origin is asked first, then he looks at whether combining or omission processes have taken place, and accordingly places the items in the various categories.

Nearly two thirds of the words in his study are composites, a category split into affixing and compounding processes. The second most productive category – shifts, which designates all words "that have been used in a new meaning and sometimes as a new part of speech" (1980:270) – he calls a "poor second" (ibid.), as its frequency is significantly lower than that of the composites. Further on, shortenings (backformation and abbreviation), loan words and blends follow in this order.

One interesting finding of his investigation is that among the composites he describes affixed forms as being more common than compounded form. And amongst the affixation processes, prefixing is more common than suffixing although smaller number of incidences of suffixing processes yield greater numbers of occurrences which leads him to claim that "English is a suffixing language" (1980:272).

5.2.1.3 Laurie Bauer: Watching English Change

Bauer excludes semantic change from his study and apart from that fact very much follows Algeo in method and approach, rendering the findings comparable. In his study, Bauer takes The Supplement to the Oxford English Dictionary which comprises the time period from 1972-1986. He samples every fifth word from each double page and comes up with 2,078 language items that occurred first after 1880. He expands the methodological view in that he also determines the first appearance of the items and categorises them according to three time periods: 1880-1913, 1914-1938, and 1939-1986; a division based on political grounds. With the help of these time slices he aims at establishing whether the productivity of the individual processes changes over time.

As the source he uses the Supplement to the Oxford English Dictionary. The language variant under investigation is General English. Bauer separates the words on the grounds of two basic questions: what sources (English or other languages) the words stem from and what "types of formations (were) used in the words coined from English resources" (1994:33).

As my view only focuses on word-formation patterns at the moment, the different source languages and the findings for the different times divisions will be left aside here.

Bauer established ten formation type categories: Abbreviations, Blends, Shortenings, Compounds, Prefixation, Suffixation, Names, Neo-Classical Compounds, Simultaneous Prefix- and Suffixation and Other.

By far the greater quantity of new words were coined by suffixation (almost half of them), followed by almost 20% resulting from compounding processes. Although different in categorisation, the figures are comparable to Algeo's study. The extent of composition processes is much higher. Bauer attributes this to "different criteria for selection of words" in the different dictionary-making processes and expresses here one weak point of the method: sampling from dictionary data is skewed by the fact that the dictionary builders do not take all words that are new in language use but choose a sub-amount of language items they consider worthy of inclusion in their dictionary.

5.2.1.4 Summarising the Categorisation Attempts

The following table summarises Caso's, Algeo's and Bauer's taxonomies with view to their chosen categories and subdivisions.

Tab. 11: Overview on formation types

Caso	Algeo	Bauer
I. Semantic Change	I. Loan Words	I. Abbreviations
II. Morphophomic Change	......Straight Loans	II. Blends
III. Functional Change	......Loans with blending	III. Shortenings
IV. Composition	......Loans with composition	IV. Compounds
......Prefixing	......Loans with shortening	V. Prefixation
......Suffixing	II. Shortenings	VI. Suffixation
......Blending	......Backformation	VII. Names
......Compounding	......Abbreviation	VIII. Neo-classical compounds
......Combining		
......Acronymy	III. Composites	
......Initializing	......Affixed forms	
	..........Suffixed forms	IX. Simultaneous prefix and suffix
	..........Prefixed forms	
V. Neologizing and Borrowing	......Compounds	
......German	..........Nouns	X. Other
......French	..........Verbs	
......Other modern languages	..........Adjectives	
......New Latin	..........Prepositional Phrase	
......Latin		
......Greek	IV. Blends	
VI. Miscellaneous	......Simple	
	......Overlapping	
	V. Shifts	
	......without change of form	
	......with change of form	
	V. Unknown etymologies	

For the Table 11 I usually took the first two levels in categorisation depth and the third only when it reflected facts already mentioned as findings of importance.

In Tabele 12 I chose to display selected processes of word-formation which allowed comparison among the three studies.

As already mentioned, the great deviation in the figures of the individual processes might be due to the fact that the database of language items used, i.e. the selection of the dictionary entries, can be considered as too heterogeneous. Furthermore, the inclusion or exclusion of semantic categories has an influence on the overall numbers. In light of this background, the huge differences for instance in the percentage of composition can be explained. And finally, the chosen language, i.e. General English versus Scientific English is assumed to have an influence on the percentages as well, as for example in the case of a higher number of neoclassical combining forms occurring in Scientific English.

Tab. 12: Selected processes in %

Formation type	**Caso**	**Algeo**	**Bauer**
Composition	36.7%	63.9%	82.1%*
• Compounding	11.8%	29.8%	18.5%
• Prefixation	2.2%	15.6%	12.4%
• Suffixation	9.8	18.5%	47.5%
• Combining/Neoclassical Combining Forms	8.1%	/	3.7%
Blends	1.7%	4.8%	2.4%
Abbreviations	3.1%	8.3%	1.3%

*(there is no explicit category of composition, the displayed percentage is the sum of compounds, prefixation, suffixation and neo-classical compounds)

As I do not aim at establishing the percentages of the single word-formation processes, the differences in the three approaches can be regarded here to be of minor importance, although they still indicate certain tendencies like a preponderance of composition processes in English. Furthermore, the weaknesses of a dictionary-based approach and its reliance on selection processes that are not within the power of the researchers but of the dictionary makers are mentioned which could be remedied by a corpus-approach. This corpus-approach could cover all word-formation phenomena in their true extent, as it is only by its representativeness that the corpus is limited.

5.2.2 The Corpus-Based Approach

Corpus-based approaches to the study of word-formation patterns and their productivity go back to the notion of a dynamic or monitor corpus as described in Sinclair (1982). A monitor corpus is a corpus where new texts are added and old ones discarded on a continuing basis. The first monitor corpus to be built was a 100 million word corpus of text from the Times dating from the early 1990s created within the AVIATOR project and described to some extent in Renouf (1993), Baayen and Renouf (1996) and Bauer and Renouf (2001). Software filters were used to search each month of the newspaper data for new words which were then compared to a master wordlist in order to determine whether the data would reveal lexical innovation and change. It is only the first in this series of articles that uses this new methodology exclusively to find categories of word-formation patterns. The later studies subsequently use the word-formation patterns as a starting point to discuss and establish their productivity.

Renouf (1993), in the early stages of the AVIATOR project, establishes eight word-formation categories based on the retrieved language data in the texts from February to March 1991 (cf. 19:281)

1. Onomatopaeic items like *plonkily*
2. Transcriptions of speech like *lurvely* (for *lovely*)
3. Puns like *last-past-the-post system*
4. Blends like *kissogram*
5. Clippings like *videograb*
6. Compounding like *sound-bites*
7. Derivation (which divide up into suffixes, combining forms like *–related* and *–friendly*, and prefixes)
8. Compounding of word strings that contain more than one hyphen like *off-the-top-of-the-head suggestions*

Although the ground on which the categorisation is based can be questioned, Renouf's examples reflect the kind of data corpus findings yield. She furthermore comments on some of the retrieved processes in terms of their frequency. Derivation and compounding are mentioned as chief sources of new language items and again within the categories certain lexico-grammatical trends are described more in detail, i.e. the word-class shifts.

Renouf's article is also the beginning of criticism of the dictionary-based investigations as the corpus data reveals their shortcomings in that language items are found that do not fit the long-established categories and word-formation theory in general. This is later on formulated in Bauer and Renouf (2001:101): "Not only do we report on patterns that are not described in the major handbooks (...) but we show that some of the patterns that are being used productively in the English of the 1990s break some principles that are laid down as absolute in some of the theoretical works."

Corpus-based approaches to lexical description of language use are an advance in the methodology of the investigation of word-formation patterns although categorisation was often only used as a starting point in their studies. Dictionary-based categories were considered as not reflecting types of word-formation right across the board and only corpus-data to reflect a true picture.

The study of the three different dictionary-based approaches and the one corpus-based were important for choosing my own sampling methodology and way of categorising word-formation processes with the aim of reflecting the situation of environmental engineering English. The approach will be described in the following paragraph.

5.2.3 Establishing Suitable Categories of Word-Formation Patterns for Environmental Engineering Data

Rapidly growing information exchange by means of mass media like TV and internet and the increasing rate of international cooperation in different specialist fields have a huge impact on the amount of new words being created and disseminated. Bauer (1994:30) says that lexical change "reflects change in society", and as environmental engineering is still a young and developing branch, we can still expect lexical innovation to continue.

Students might attribute the fact that they encounter new words to the fact that they are language learners but they need to be made aware of the notion that language is dynamic and new words are produced while old ones fall out of use. In addition to understanding the meaning of lexical items in their context, as discussed in the previous chapter, their understanding of morphological information is another vital key to making sense of learned language. Once they are familiar with certain base words, derived forms will be easier to grasp. They need to know which word-formation processes are active and how the resulting new language items differ from the bases in terms of a different lexical meaning and added syntactic information, as produced by the addition of affixes.

In order to find out which processes of word-formation are active in environmental engineering English, corpus findings can be consulted. I will look at the question of whether Caso's claim that an "explosion of scientific knowledge is having a considerable linguistic effect as the need to communicate new concepts forces the devising of new terms" (1980:101) holds true for the specialist language of environmental engineering English as well.

As the ENGICOR corpus has not been designed as a monitor corpus, the applied methodology in the present study must be different to the discussed approach. Furthermore, the focus will not be on establishing percentages for the frequency of the individual processes but to find out which processes are characteristic of and active in environmental engineering English.

Every word-formation pattern found in the corpus was looked at in the light of the following basic questions:

1.) What elements join or are involved in the word-formation process?

2.) What are the results of the process in terms of form and meaning(s)?

Where it is appropriate, a third question will be added:

3.) In which word classes does the process happen?

This question will only be dealt with if the relevant word-formation process occurs with significantly higher frequency within a certain word class.

The subjects of investigation were indirectly derived from the corpus. Here the usage profiles of the 500 keywords come into play again. In their second part, corpus-attested word-formation patterns were collected during the phase of the empirical collection of the corpus information.

In a second step, all of the attested word-formation patterns were assigned to their different classes, like affixation, compounding, etc. Given the nature of the usage profiling data, this could be done within the different word class categories, like adjectives, nouns, verbs, etc. Although this method employs the usage profile data as a mediator, the morphological information can still be regarded as corpus-derived, as no further categorising has taken place than the simple decision concerning the usage profile sheets stating whether the resulting language items are to be regarded as derived from a compounding or having a derivational relationship with the concerned keyword.

Once all language items from the usage profiles were determined as belonging to one specific word-formation process, I tried to put the investigated, corpus-derived information from the processes in a certain order. To this end and roughly modelled on Algeo's views (1980), the ordering element in my categorisation attempt was the question as to whether there has been a combination of word-formation elements or whether there was an omission of word-formation elements.

A combination of word-formation elements (marked with "+combination") comprises compounding, combining and derivational processes. An omission of (a) word-formation element(s) results in abbreviations/initialisms, clippings and backformation (marked with "+omission"). If there was an omission and a combination of elements the category applies to blends (marked with "+omission/+combination"). And if neither a combination nor an omission has taken place the category comprises language items that result from conversion processes (marked with "-omission/-combination").

Another category collects items that do not comply to the described scheme; word manufacture instances and loan words will be collected here. And a last category will describe instances of semantic change.

The systematic approach of consulting the usage profile information as mentioned above can only account for compounding, combining and derivational processes. Corpus tools are best suited for searching instances and establishing the frequencies of these processes. Still, the comprehensive work with the corpus has brought examples for the remaining processes to light, be it in the usage profile or unearthed by chance in the context of other words. So the claim that all examples are corpus-derived can still be maintained.

All processes that will be discussed are form-oriented. Only the category of semantic change is not form–oriented, though later I will prove that the co-text of new language items produced by means of semantic change will reveal enough form-based information to be sure that the meaning of the word is different from its use in General English.

The described categorisation attempt can be summarised in schematic format as follows:

1. + Combination:
- compounding processes
- combining processes
- derivational processes

2. + Omission:
- abbreviations/initialisms
- clippings
- backformation

3. + Combination/ + Omission:
- blends

4. - Combination/ - Omission:
- conversion

5. Others:
- word manufacture
- loan words

6. Semantic change

Where appropriate, these processes will be looked at in the light of the above-mentioned questions. Concerning their meaning, information from standard works of morphology will be consulted in order to compare with the corpus data.

5.2.4 + Combination

5.2.4.1 Compounding Processes

5.2.4.1.1 Introduction

Compounds are complex lexical items "consisting of more than one base and functioning both grammatically and semantically as a single word" (Quirk 1985:1567). As mentioned in the categorisation attempts above, compounding patterns contribute a great deal to the extension of the lexicon in General English and, as will be proven, in domain-specific language as well.

Discussing their function in a specialist language, Sager (1997:30) remarks: "Functionally, (derivation and) compounding serve the purpose of closer determination of a concept - narrowing its intension – while at the same time showing the relation that exists between the new concept and its origin." Especially in such a very specialised branch of engineering, it is of paramount importance to designate objects with highly specific meanings and characterise process parameters by assigning them a suitable name. Consequently, environmental engineering English is assumed to be a good source for domain-specific compounds. It will furthermore be interesting to investigate which word-formation patterns these compounds follow and whether this displays domain-specificity as well.
Compounds, in general, can occur as a single word, as a hyphenated word, or as (at least) two words. Most compounds have the main stress on the first syllable and occur in a modifier-head structure where the first element describes the second, and the second being the one that determines the main characteristics, i.e. word-class, of the resulting word form. Furthermore, many compounds are hyponymic, that is they describe an object or an entity that is a hyponym of its head and are thus subordinate. With such a variety of parameters, it should be easy to distinguish between real compounds and syntactic construction from the vast amount of two and multiword combinations in a corpus. However, there are many exceptions to these rules. Bauer (1998) discusses the question "When is a sequence of two nouns a compound in English?" and verifies amongst others the above-mentioned characteristics. He comes to the conclusion that "there is no strong evidence for a distinction" (1998:85), thus there is no surefire way of classifying a two or multiword item as a compound or syntactic construction.

With this in mind, the current study will take a pragmatic stance looking for more typical word combinations that designate concepts in environmental engineering and leave problematic cases aside. This is done because the problem whether a two or multiword combination is really a compound is not relevant to this study as it focuses on finding data on which language teaching can be based. It will moreover be examined how the lexical elements typically combine to form new meanings.

In terms of classifying the compounds, various methods have been used. Bauer (1983:201) claims that the "normal way of classifying compounds is by the function they play in the sentence as nouns, verbs, adjectives, etc". He as well as Marchand (1969) uses this type of classification. Other attempts used the syntactic functions the involved elements could take if the compound was expressed in a short defining or describing phrase (cf. Quirk 1985). Classifications on grounds of semantic classes are problematic as there are arbitrary ways to establish different semantic categories (cf. Plag 2003:142).

I opted for a classification that considers the compound-forming elements as well as the compound itself as a result in terms of their word classes. Although it is admittedly difficult with some words to decide which word class they belong to I preferred this approach over the arbitrariness of assigning semantic classes to the found items. Word form classes are at least a set of categories that are generally agreed upon whereas the number of semantic classes varies according to the observers and their arbitrary decision making. Using a corpus approach though, brings certain common semantic features to light which I will discuss as well (cf. p. 108ff.).

In the following overview on compounding patterns in environmental engineering English I will restrict myself to two word combinations only. They yield the most instances and clearer patterns. Multiword compounds will be discussed later.

5.2.4.1.2 Categorising Compounding Patterns

In order to elicit the existing word-formation patterns in environmental engineering English, the usage profiles of the first 500 keywords were searched. Regarding the compounding patterns, these usage profiles provided enough word material to illustrate all common processes, especially the noun+noun, adjective+noun patterns, etc. In rarer cases like compounds involving particles, the usage profiles did not yield enough information and potential elements of these words were searched in a concordance to check whether these types of word-formation patterns were in the corpus at all.

For the following classification of compounding patterns only two-word combinations were chosen. At first sight, the established compounds were classified according to the nature of their elements, whereas on further examination the resultant compound is looked at. If applicable, it will be mention of the usage profile category, i.e. whether the pattern is also found amongst abbreviations or chemical nouns.

It was found that the elements used to form English compounds in this specialist language were nouns, adjective, verbs, particles and adverbs. Furthermore, it can be said that the patterns are not all used to the same extent. Nouns play an outstanding role in the formation of new words and noun+noun combinations are by far the most common type of compound (cf. Bauer/Huddleston 2002:1647). As this statement gets confirmation from many researchers (Bauer 1983:202, Plag 2003:145), I did not aim at a frequency analysis of this fact. A look at the usage profiles' N+1 and N-1 sections is enough to get an idea of how frequent they are as these sections consist of a huge amount of noun+noun compounds.

5.2.4.1.1 Noun-Centred Compounds

Type 1a	**noun + noun**
	filtration plant, soil erosion, ash content, infiltration sump, asphalt binder, LFG collection, effluent nitrate, slurry-phase
results in	compound nouns, sometimes in attributive position

Among the very productive noun compounds, combinations of two nouns are by far the most productive. Most combinations display a modifier-head structure in order to assign differentiated meanings to objects and entities with special functions in science and technology. Exocentric compounds which do not sport this right-headedness do not play a role in ENGICOR, and are thus not discussed here. There were also no copulative compounds found by the mentioned methods. So the noun+noun compounds seem to be at least in this regard a huge uniform mass of nouns in two-word combinations. Chemical nouns appeared to behave in the same way: they were found in modifier position (*ammonia addition, manganese level*), in head position (*effluent nitrate*), or as both constituents (*aluminium chloride, carbon hydroxide*). Abbreviations compound with nouns resulting in constructions like *AMD production* or *BOD level* with the abbreviation in the modifier position and *wastewater COD* and *soil pH* where the abbreviation is in head position.

Noun+noun compounds result in compound nouns which can also be used in attributive position in order to describe another head noun and thus create multiword compounds (i.e. *slurry-phase bioreactor, froth flotation circuit*).

The semantics of noun+noun compounds

Compounds in domain-specific language like *curbside program, barrier performance* and *GAC beds* are not easily understood without having encountered them before in the situational and linguistic context in which they occur. That means that they must be learnt even by native speakers. Others can be easily understood in context (*bioremediation process, soil erosion*). Adams (2001:86) comments on that phenomenon as follows: "As well as expressing a new lexical concept, a compound serves as a compact version of a longer expression in its neighbourhood which furnishes a full gloss of it."

In the consulted works of word-formation, the compounds have often been subject to classifications according to either semantic features or syntactic and logic relations. Just as an example, Quirk (1985:1573-74), amongst others, mentions verbless compounds of the type "subject and object" and differentiates between compounds that follow the example of *windmill* where the first noun powers or operates the second, followed by many others of the same style. Marchand differentiates noun+noun compounds according to whether they follow different types like *steamboat, policeman*, etc. Adams (2001:83 ff.) starts describing a few categories, and claims that compounds can be classified into types like "the head may indicate a location of an activity or entity", "the modifier indicates a cause", etc. All these attempts of classifying are based on the fact that the elements of a compound have different relations to each other. Adams comments on her attempt at classification as follows: "like many others, (it) relies on rather loosely defined distinctions, and these groupings may not seem particularly homogeneous" (2001:84). Despite all of the above-mentioned elaborated classification attempts a common classification has not yet been devised as the meaning of compounds depend on a variety of multi-facetted factors. Plag (2003:151) sums this up as follows: "..., the interpretation of noun-noun compounds is highly variable and depends on the argument structure of the head, the semantics of the two nouns, the possible conceptual relationship between the nouns, and on the surrounding discourse."

The motivation to nevertheless look at the semantics of the noun compounds in environmental engineering English does not come from investigations into General English but from an article on term formation by Sager (1997:30): "According to the nature of the nucleus," (i.e. head) "it is possible to distinguish between types of compounds that designate objects, properties, processes, or operations". In order to find out whether this is true for compounds in ENGICOR, the usage profiles were searched again for evidence of these categories. Broadly speaking, Sager's categories were reflected in ENGICOR but according to the specificity of the language material in that specific corpus and also depending on the collected text types (reports, summaries, technology fact sheets, textbook, internet journal), more specific categories could be established. Looking at the two word compounds in environmental engineering English, the majority of items could be assigned to one of the following semantic categories:

Tab. 13: semantic categories in Environmental engineering English

processes/operations	process indicators	process agents/ products	process equipment/devices	process locations
dam construction	concentration level	foundry waste	clay cyclone	combustion chamber
DNAPL contamination	moisture conditions	exhaust gas	construction equipment	collection point
dust suppression	contaminant mobility	carbon compounds	collection sump	batch kiln
VOC emission	data quality	demolition debris	distillation columns	aeration basin
soil erosion	bulk density	metal contaminant	concrete paver	bottle depot
membrane filtration	soil depth	wastewater constituents	retention dam	clay pond
lime addition	removal efficiency	asphalt binder	gas chromatography	landfill bioreactor
bottle resterilization	groundwater pH	biosolids compost	HDPE liner	concentration unit
data collection	inlet pressure	lime slurry	aeration device	contaminant plume
aquifer cleanup	material size	concrete mix	air compressor	hammer mill
ash removal	material composition	construction fill	subsurface barrier	adsorption site
bacteria inactivation	bicarbonate alkalinity	landfill leachate	drainage aggregate	bedrock aquifer
biodegradation process		scrap iron	infiltration sump	asphalt plant

These five semantic categories mirror the immediate reality of environmental engineering. Processes happen under certain conditions (described by process indicators) with the help of certain equipment in specific locations. Agents are involved in these processes and they result in certain products. It is strongly assumed that these categories are topic-dependent as Sager's categories are much more general, and text-type-dependent too. It might be that with other text-types animate agents would play a bigger role or with more instructional text-types tools would come into play more. Still, language use reflects very well the restricted world of environmental engineering. I tried to illustrate this in the Figure 11 below:

Fig. 11: semantic categories diagram

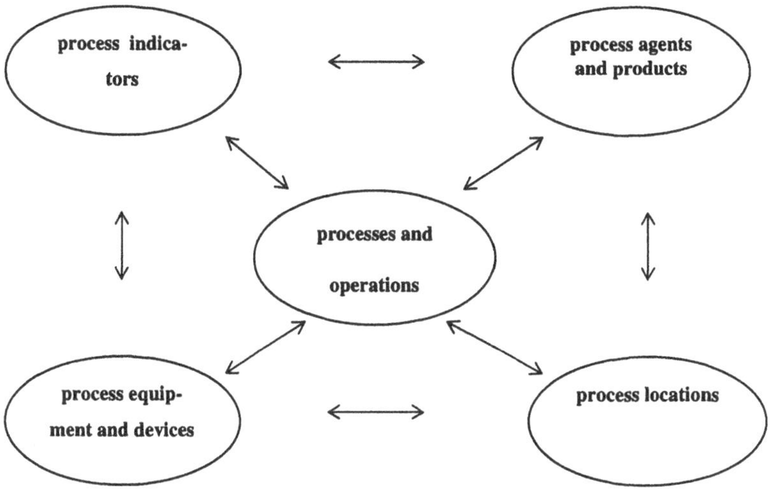

Type 1b	noun + verbal noun
	curbside recycling, hypochloride bleaching, LFG monitoring, cosolvent flushing, coal mining, copper leaching, toxicity testing
results in	compound nouns, sometimes in attributive position

This type comprises compounds that consist of a noun as a first component and a verbal noun as second component (*curbside recycling, cosolvent flushing*). First elements can be abbreviations (*LFG monitoring*) or chemical nouns (*hypochloride bleaching*) as well.

What becomes obvious from looking at the corpus examples of this compound type is that in terms of their semantics they only describe processes and operations; a specific detail that derives from their character as verb-derived nouns.

Like type 1a, compounds of this class form compound nouns and can occur in attributive position as well.

Type 1c	verbal noun + noun
	composting facility, flushing rate, sampling period, mining location, settling pond, leaching field, drinking water
results in	compound nouns

In this type a verbal noun combines with a noun to finally form a compound noun. This category is related in form to type 4. The verbal nouns serve the specific purpose of describing the head noun and distinguish them from the more general hyperonym (settling pond < > pond).

Type 2	**verb + noun**
	leach field, drainfield, spill control, shrink-wrap, turntable, washwater, borehole
results in	compound nouns, sometimes in attributive position

In the second category the first element is a verb and the second element a noun. Two main statements concerning verb+noun combinations hold true for the later-discussed examples of combinations involving verbs as well: first, the process of producing compounds involving verbs is only marginally productive in English and second, there is great uncertainty as to whether the concerned verbal component is really a verb or in fact a noun, as many words belong to both categories and can – in combinations – not entirely be distinguished. The first statement can be emphasised by Adams (2001:101) saying that "(i)n a striking contrast to noun and adjective compounds, verb compounds are very uncommon". Support for the second statement is found in the Bauer/Huddleston (2002:1651): " A great many lexical bases in English can be either verbs or nouns, and as a result there may be uncertainty or indeterminacy as to whether a component of a compound is one or the other." ENGICOR provides the following examples for that: in *spill control* or *drainfield* we cannot 100 per cent sure whether the first element is a verb or a noun.

The resulting word combination is a compound noun, sometimes also used in attributive position (i.e. *turntable model*). Combinations of verbs and verbal nouns are not attested.

Type 3a	**adjective + noun**
	dry mix, hazardous waste, hydraulic backfill, greenfields, raw material, drycleaners, full-scale
results in	compound nouns, sometimes in attributive position

In type 3a a first element adjective combines with a second element noun in order to form compound nouns. Again there are problems to assign many word combinations to this class. This time the uncertainty stems from the fact that the adjective+noun combinations are sometimes not clearly classifiable as a compound and can be regarded as a syntactic construction as well (i.e. *hydraulic backfill, hazardous waste*). In these examples the first element stress rule which prevails with most compounds does not fit them as in more obvious examples like *dry mix* or *raw material* (cf. Adams 2001:81). However, on grounds of their widespread occurrence and domain-specificity they are included in this class here as well. Members of this class can also be used in attributive position (i.e. *full-scale treatment*). Adams (2001:81) experiences a similar uncertainty: "… we have no firm criteria separating compound and phrase". She argues that if a modifier were to be added to her examples (i.e. a *guided missile, a flying saucer*, etc.) as it can be done with any syntactic construction of the type adjective noun, the language items need to be reanalysed in terms of their meaning (i.e. *a [carefully guided] missile, a [high flying] saucer*). I argue that the same would happen to two-word items like *hydraulic backfill – extremely hydraulic backfill*, and *hazardous waste – very hazardous waste*. These adjective+noun combinations have acquired special status on grounds of their use in the domain-specific language and are therefore considered here as compounds. I strongly believe that combinations like *solid waste* or *hazardous waste* are of another quality, as are *acidic waste, biodegradable waste* or *dangerous waste*. In the case of the example *hydraulic backfill*, it is clearly different from *special backfill*. This idea can be backed up by the frequency of co-occurrence: *solid waste, hazardous waste* or *hydraulic backfill* are the adjective+noun collocations which are the most frequent in a concordance (8-10% of the total occurrences of *waste* or *backfill*). Adams comments on this phenomenon as follows: "The adjective in an adjective+noun sequence often signals a characteristic property in contrast to its 'temporary' significance elsewhere:…" (Adams 2001:81). She adds: "(r)elated to this is the susceptibility of attributive adjective-noun collocations to lexicalization ( ), which involves a loss of independence in the constituents,…" (ibid.). Thus they become frequently used, set terms in the special language.

Type 3b	**adjective + verbal noun**
	hydraulic jacking, reactive blending, tensile testing, dry cleaning
results in	compound nouns, sometimes in attributive position

This type is similar to the previous but has a verbal noun as second element which again assigned process character to the resulting compound (compare type 1b). The process results in compound nouns.

Type 4	**verb+-ing + noun**
	oxidizing agent, drying bed, clarifying tank, confining layer, denitrifying sludge, dewatering chemicals
results in	compound nouns

Everything that was said for type 3 in terms of compound noun status applies to 4 as well. In appearance it is also similar to type 1c but although both first-element constituents have the suffix *–ing*, it is a verbal noun in type 1c and a participial adjective (from the present participle) in this type. The second element is a noun which is characterised by the participial adjective. The first elements describe an active quality of the nouns. It is obvious that the resulting compound is an agent or a device of some kind that changes the properties of something: an *oxidizing agent* is an agent that oxidises something; a *clarifying tank* is a tank that clarifies something. An example from General English can also be found in the corpus – *developing countries* – in this example a reflexive meaning is expressed: the country develops itself. *Driving force* and *leading edge* are other examples of that kind that have become lexicalised. The results of this compounding process are compound noun.

Type 5a	**verb+-ed + noun**
	activated sludge, mixed plastics, chlorinated solvents, constructed wetlands
results in	compound nouns, sometimes in attributive position

Type 5a consists of a participial adjective (from the past participle) and a noun. Here again the frequency and central significance of the chosen examples made me include these language items under the heading of compounds. By means of the participial adjectives, the quality of the nouns is described more in detail and so they become hyponymic. *Activated sludge* and *chlorinated solvents* are items which are central to the field of environmental engineering. The results are compound nouns.

Type 5b	**verb+-ed + verbal noun**
	automated sorting, layered co-mingling, accelerated aging
results in	compound nouns, sometimes in attributive position

Combinations of adjectives and verbal nouns are extremely rare and it remains questionable whether this type should be maintained as a class at all. It is just the status of the found examples (*accelerated aging, automated sorting*, and *layered co-mingling*) which are commonly known and used as rather stable combinations that provide arguments for maintaining this type. The results of the process are again compound nouns.

Type 6a	**particle + noun**
	backpressure, aftertreatment, byproducts, in-depth, on-dam, off-site, downslope
results in	compound nouns, compounds adjectives, compound adverbs

In this type a particle, i.e. a preposition, combines with a noun to form a new language item. According to Bauer "(t)his is a very productive pattern" (1983:206). In *backpressure, aftertreatment* and *by-products*, the elements combine to form a new compound noun, where the preposition serves as an attribute describing the noun in a common modifier-head structure. Constructions like *in-depth, on-dam* or *off-site* go back to prepositional phrases (*to discuss something in depth, something is situated on the dam, something is disposed or transferred off the site*) and must be product of a restructuring of these phrases to particle-noun constructions that serve attributive purposes as they result in compound adjectives and compound adverbs (*off-site disposal, re-deposited further downslope*).

Type 6b	particle + verbal noun
	backwashing, upgrading, upsizing, off-gassing, after-processing
results in	compound nouns, compounds adjectives

In this type a particle seems to combine with a verbal noun. But most cases can be assumed to be derivations from the belonging verbs or process-oriented constructions from the belonging noun (*backwashing* from *backwash v/n, upgrading* from *to upgrade/the upgrade, upsizing* from *to upsize/size-up* and *off-gassing* from *to gas off/offgas)*. Of the examples *after-processing* alone seems to follow the same parallel as *aftertreatment*. Whereas with *backwashing, upgrading*, etc. the result is a noun, *after-processing* is used in attributive position to describe another noun.

5.2.4.1.2 Verb-Centred Compounds

The word combinations that are described in the next three categories consist of a verbal head element with either a noun, another verb or an adjective in the modifying position. The striking feature of assumed verb compounds is that they are only compounds regarding their form. Looking at their origin, they very often stem from conversion or backformation processes, or are formed on the basis of analogy processes. Literature discussing verb-compounds agrees that neither real verb-compounding in particular nor the other processes generating complex verbs are very frequent (Adams 2001:100, Bauer 1893:208). Adams furthermore comments on the effects on the participants in discourse: "Seen out of context, a verb compound may be judged doubtfully acceptable, or even ill-formed, and when encountered in context, it is likely to strike the hearer or reader as rather conspicuous" (2001:101).

It is as well Adams who claims that the motivation to nevertheless use verb compounds and thus accept their peculiarities is the "need to background known information, to allow a constituent to occupy a position of salience or to narrow down the meaning of the verb." (2001:109) So we have the *dry* in *dry clean*, the *size* in *size-reduce* and the *back* in *backfill* that describes the verbal heads more in detail, a function characteristic for a special language where precision and shortness are important.

Type 7	**noun + verb**
	field-test, size-reduce, landapply
results in	compound verbs

For this type a noun and a verb combine to form complex verbs. None of the corpus-attested examples are genuine compounds but rather results of conversion or backformation processes, *field-test* from the compound verbal noun *field-test* and *size-reduce* from the compound noun *size-reduction*. This type at least seems to be an area of modest productivity (cf. Bauer 1983:208) with recent General English examples like *blockbust* and *carbon-date*. According to Bauer (ibid.) it is not just backformation but also conversion that produces these results (like in the case of *carbon-date*).

Type 8	**verb + verb**
	starve-feed, push-pull, freeze-dry
results in	compound verb, compound adjective

This type consists of two verb elements which form an assumed compound verb occurring in the corpus very often in attributive position in front of nouns (*freeze-dried bacteria, push-pull test*)

Bauer (1983:208) says about this type that it is "exceedingly rare", a statement that is confirmed by the rare instances of corpus findings. Moreover, two of the three retrieved examples cannot count as a real compounding process, *starve-feed* is a backformation from the longer existing *starve-feeding* and *freeze-dry* from *freeze-drying* (cf. Bauer/Huddleston 2002:1660).

Push-pull seems to be a genuine coordinative verb compound where both meanings apply at the same time.

Type 9	**adjective + verb**
	dry-grind
results in	compound adjectives

Here an adjective is found in the modifier and a verb in the head position. All the consulted literature yielded just one mention of this type. Bauer (1983:212) gives examples like *high-rise (tower)* and *quick-change (artiste).* The corpus findings can only add one more example of this kind: *dry-grind (tests)* which confirms the pattern of use being in attributive position in front of a noun. Bauer furthermore points to a peculiarity of this type which is that concerning the form, the first element is undoubtedly an adjective which nevertheless functions semantically as an adverb (cf. Bauer 1983:212); a circumstance which might have hindered its productivity.

As there exists a verbal noun *dry-grinding*, its closeness to *freeze-drying* and *starve-feeding* allow me to conclude that it is not a result of a compounding process but of a backformation.

Type 10	**particle + verb**
	backfill, uptake, by-pass, out-perform, out-compete
results in	compound verbs

Type 10 includes combinations of particles and in this case verbal elements which are in many cases only at first sight compounds. Some like *backfill* are conversions from the noun by the same form, others like *by-pass* are products of inversion processes of phrasal verbs (cf. Adams 2001:72). A third type can be represented by examples like *out-perform* and *out-compete*, *over-* and *under-achieve*. Whereas the latter examples are accepted as real particle-noun compounds, the status of formations with *out-* remain disputable. Bauer (1983:208) calls them "genuine verbal formations", Plag however argues, that as *over-* and *under-* keep the same meaning in the compounds as they have in their free forms, *out-* acquires the meaning of going beyond a limit (cf. Adams 2001:73) and is thus no longer a compounding element. Adams furthermore comments: "Certain initial particles, semantically distinct from their independent homonyms, occur productively in a series of items and are then even more like prefixes." (Adams 2001:71). As the figurative meaning of *out-*, given in the OED comprises "beyond the limits", I have chosen to treat *out-*+verbal element as compounds as well.

It is particularly their being so different from the usual left-headed verb + particles construction that accounts for markedness.

The results of the described processes are verbs.

5.2.4.1.3 Adjective-Centred Compounds

Compounding processes to form complex adjectives are manifold and those involving a noun in the first position are particularly productive. Other first elements can be adjectives, adverbs and particles. In second position we find either normal adjectives or participial adjectives.

Type 11a	noun + adjective
	zone-dependent, vapor-permeable, ozone-reactive, debris-free, landfill-specific, airborne, operator-friendly, clay-rich
results in	compound adjectives

Having said that the processes forming complex adjectives in general are very productive, the compounding of nouns and adjectives count as the most frequent. The nouns in first position correspond to the complements in prepositional phrases (*pollution-sensitive – sensitive to pollution, debris-free – free from debris, clay-rich – rich in clay*). Those classes of semantic categories as established by some authors – comparative or intensifying function like in *bone-dry* (cf. Bauer/Huddleston 2002:1656), measure terms like *ankle deep* (Hansen 1990:60.) do not play a role in environmental engineering English. One phenomenon, however, applies to the domain-specific language in particular. This is the property of some elements to "appear in series of forms" (Adams 2001:96), like *–blind, conscious*, and *–mad*. Environmental engineering English though has its own series-forming second elements like *–sensitive, -specific, -free, -dependent, -intensive, -rich, -prone, -resistant, -tolerant, -inhibitive*, and *–borne*. The task of the resulting complex adjectives is above all to describe properties and characteristics of materials, products and objects (*UV-sensitive chemical, corrosion-resistant alloys, lead-free solders, airborne particles, herbicide-tolerant plants*).

A fascinating consequence of the fact that many of the second elements are series-forming is that the single elements show a clear affinity with a certain semantic preference and moreover semantic prosody, a phenomenon that can be made visible by concordancing (see Table 15).

Tab. 14: pattern: series-forming noun-adjective+noun (phrase)

second element	attested collocations	semantic preference	semantic prosody
-free	debris-free product lead-free solders mercury-free equipment plastic-free organics	products, devices or substances that are free from a contaminant	positive

-rich	clay-rich tailings fuel-rich zones iron-rich barren solution ozone-rich gas soil-rich area grass-rich yardwaste	locations or sub-stances that are rich in a certain element or substance	positive
-tolerant	drought-tolerant vegetation herbicide-tolerant plant metal-tolerant plants	plants that are tolerant towards certain con-taminants or climatic conditions	positive
-sensitive	redox-sensitive contami-nants UV-sensitive chemical pollution-sensitive mayfly	substances or animals that are sensitive to certain influences (contaminants, chemical reactions, climatic conditions)	positive or negative depending on whether the effect is desired or not
-prone	erosion-prone silt erosion-prone soil strike-prone area	location that is prone to a negative influ-ence	negative
-intensive	area-intensive construction capital-intensive system labor-intensive task energy-intensive industries	an endeavour that is intensive in input	negative
-resistant	abrasion-resistant materials corrosion-resistant alloys fire-resistant lubricants water-resistant surface	material that is resis-tant to chemical or mechanical influences	positive
-borne	airborne concentrations airborne dust fluidborne species soilborne plant diseases wastewater-borne nitrogen waterborne diseases waterborne pathogens windborne bacteria	harmful substances or diseases carried by a certain medium	negative
-friendly	userfriendly practices userfriendly flow chart userfriendly graphics maintenance-friendly mate-rials environment-friendly toothbrush	an object, entity or practice which is ei-ther suitable or not harmful	positive

The most frequent second element *–specific* displays a rather neutral semantic prosody and the semantic preference is also a more abstract one pertaining to a piece of information, a material or an entity specific to or for a certain location, substance or process (compare: *application-specific information, facility-specific information, ion-specific electrode, project-specific answers, site-specific crite-*

ria, site-specific materials, waste-specific parameters). The use of the compound adjective *site-specific* is – by the way – significantly more frequent as in General English and thus occurs amongst the first 500 keywords of environmental engineering English established in the comparison between ENGICOR and the BNC.

Type 11b	**noun + verb+-ing**
	iron-oxidizing, colloid-mobilizing, methane-forming, iron-containing, groundwater-bearing, oil-absorbing, odor-causing
results in	compound adjectives

In type 11b nouns appear in first position and present participial adjectives in second. If turned into a syntactic phrase, the noun corresponds to the object of the underlying verb.

As this type of syntactic construction is quasi-unlimited in language, so is the word-formation pattern producing compound adjectives of this kind. In the corpus though there are fewer and less varied examples than for type 11a.

With this type again, there are some second elements which are more frequently used and thus appear in a series. They are also typical for environmental engineering English and will therefore be shown in the following table together with their semantic preference and semantic prosody.

Tab. 15: series-forming noun-present participial + noun (phrase)

second element	**attested collocations**	**semantic preference**	**semantic prosody**
-containing	asbestos-containing soils bromide-containing waters carbon-containing materials chlorine-containing organics iron-containing zones oxygen-containing gas sulfide-containing rock virus-containing groundwater clopyralid-containing feedstocks	substance or location holding a chemical or another kind of impurity	negative only if chemical is a contaminant
-bearing	dioxin-bearing wastes copper-bearing rock fish-bearing streams groundwater-bearing unit metal-bearing solutions sulfide-bearing minerals	very close to *–containing* but there is an element of not just holding the chemical substance or the objects but of	either neutral, positive or negative according to the wider context

	water-bearing zones chromium-bearing waste-water	producing them as well	
-forming	anion-forming elements chlorate-forming reactions glass-forming process methane-forming bacteria rock-forming processes smog-forming chemicals acid-forming mine waste	processes, substances or living agents that produce a chemical or some other kind of result	either neutral, positive or negative according to the wider context
-generating	acid-generating mines electricity-generating stations metal-generating tailings power-generating facilities revenue-generating activities waste-generating companies acid-generating waste rock	facilities or processes that produce an outcome	either neutral, positive or negative according to the wider context
-causing	allergy-causing pollens erosion-causing events odor-causing compounds taste-causing compounds disease-causing microorganisms	living agents or activities that cause a negative effect	negative
-oxidizing	iron-oxidizing bacteria methane-oxidizing bacteria	living agents that change the properties of a substance	positive for these two examples

Similar to the adjectives produced by type 11a, 11b-type compound adjectives mostly pertain to either processes and properties in environmental chemistry, here with an active meaning due to the involvement of present participial adjectives.

Type 11c	**noun + verb+-ed**
	air-saturated, bacterial-laden, cement-based, metal-enhanced, sewage-contaminated, soil-related, lab-generated
results in	compound adjectives

Here nouns combine again this time with past participial adjectives to form compound adjectives. If paraphrased by a syntactic construction, the nouns mostly correspond to the passive of a participial phrase (*site-related – related to the site, sewage-derived – derived from sewage*). Typical Environmental-Engineering English-specific compound adjectives and their semantic preference and semantic prosody are illustrated in the table below. Compounds with –*based* as second element are particularly frequent.

Tab. 16: series-forming noun-past participial + noun (phrase)

second element	**attested collocations**	**semantic preference**	**semantic prosody**
-related	combustion-related emissions composting-related program gas-related problems mining-related sites sediment-related concerns waste-related activities water-related factor	two entities that have a certain relationship, *-related* carries less meaning load than the more specific words elements like *–laden*, is open for more constructions	neutral
-based	chlorine-based cleaners coconut-based GAC computer-based technology lead-based paint Jersey-based firm Ontario-based systems lime-based fixation market-based approach vegetable-based fluids technology-based limits paper-based hydromulch	either a product based on a certain substance, or a facility based in a certain location, or a technology, approach or similar based on certain facts	neutral
-contaminated	DNAPL-contaminated aquifer lead-contaminated soil pesticide-contaminated sites sewage-contaminated zone metal-contaminated biomass	object or location contaminated by a contaminant	negative
-laden	bacteria-laden water cavity-laden surface copper-laden solution metal-laden foliage solids-laden water VOC-laden fumes BOD-laden wastewater	substance full of contaminants or intrusives of some kind	neutral to negative
-treated	herbicide-treated pasture iron-treated zeolites pressure-treated wood UV-treated wastewater samples	objects treated by a chemical substance or a physical phenomenon	neutral
-derived	sewage-derived contaminants plant-derived materials	product derived from a substance or a facility	either neutral, positive or negative according to the wider

	refuse-derived fuel waste-derived compost products landfill-derived methane gas		context
-generated	combustion-generated particles contractor-generated incident lab-generated data sunlight-generated heat penetration computer-generated light source	result caused by a process or an object	either neutral, positive or negative according to the wider context
-filled	permanganate-filled fracture air-filled box fibreglass-filled strapping iron-filled fractures water-filled system wood-filled plastics	location or device filled with a substance	neutral
-induced	oxidation-induced changes process-induced effects recharge-induced sinking tracer-induced perturbations water-induced erosion wind-induced turbulence	a change in state caused by a process, substance or phenomenon	negative

Type 12a	**adjective + adjective**
	anaerobic-aerobic, organic-inorganic, wet-mechanical, wet-dry
results in	compound adjectives

What the next three types have in common is that they have an adjective (also in the form of a participial adjective) as second element and produce a compound adjective. The productivity of these patterns is much weaker than of the compound adjectives with nouns in the first position. In type 12a the first element too is an adjective. This results in a copulative adjective compound where the two involved adjectives are in an appositional relationship as they both apply at the same time (*anaerobic-aerobic process, inorganic-organic mixtures*).

An example of adjective-adjective compounds with a determinative relationship is *dry-mechanical presorting* and there is also an example of *wet-mechanical preparation*. They are all quite infrequent.

Type 12b	adjective + verb+-ing
	foul smelling, nice-looking, close-growing, free-swimming
results in	compound adjectives

Real examples of that kind would be *foul-smelling* or *nice-looking*; an adjective-participial adjective compounding process involving verbs of perception as source for the second elements, which are not found in ENGICOR. Examples like *close-growing grass* or *free-swimming organisms* are treated under type 13b because their first elements – although they could be either an adjective or an adverb – they will be treated as adverb-participial adjective compounds on grounds of their underlying syntactic relationship.

Type 12c	adjective + verb+-ed
	coarse-grained, coarse-textured, fine-grained materials, dry-blended, bacterial-based, anionic-based
results in	compound adjectives

Type 12c can only be labelled a compound made up of an adjective and a past participial adjective according to its form. Adams (2001:94) says about them that they can rather be "explained as adverbial without the ly-suffix". That means that according to their syntactic function they fulfil the task of an adverb. Still, the examples cannot be put down to mistakes. From the examples it is obvious that they all follow the same pattern (*coarse-grained materials, large-grained soil, fine-textured layer, bacterial-based catalyst system, punitive-based enforcement*). Sometimes they include an element of comparison (*coarser-grained layers, finer-grained tailings*).

Type 13a	adverb + adjective
	now-soluble, now-defunct, near-ambient, near-neutral, near-impossible, partially-penetrating
Type 13b	abverb + verb+-ing
	ever-increasing, far-reaching, fast-growing, naturally-occurring, forward-looking, close-growing, deep-lying, free-draining, free-swimming, high-paying, slow-growing, cold-starting, long-standing
Type 13c	adverb + verb+-ed
	now-departed, biologically-mediated, chemically-bound, freshly-fallen
results in	compound adjectives

With the examples of these types it is just their position as compound adjectives which justifies their inclusion in the table of compounding patterns. In these types it is an adverb in modifier position and an adjective, present participial or past participial adjective that combines to form complex adjectives. They usually have a syntactic construction equivalent when used in another position in the sentence (*partially-penetrating – penetrates partially, now-defunct – is defunct now, far-reaching – reaches far, chemically-bound – binds chemically*) which can be considered a collocation, but in the form of compound adjectives some of them have reached far more conventionalised status (*ever-increasing, forward-looking* in General English and *naturally-occurring* in environmental engineering English).

Probably modelled on the established examples, we can find more ad hoc examples of type 13b like *close-growing grass* and *free-swimming organisms*. They are extremely rare though and some like *cold-starting* and *longstanding* are conversion from the equivalent nouns.

Type 14a	**particle + adjective**
	overripe
Type 14b	**particle + verb+-ing**
	overacting, outstanding
Type 14c	**particle + verb+-ed**
	back-calculated, over-applied, outdated, updated, outlived, outperformed
results in	compound (participial) adjectives

The insufficient quantity of data does not allow for an own analysis of the pattern. Following the form criterion first, the language items are built by an initial particle followed by an adjective, a present participial adjective or a past participial adjective. As long as the meaning of the particle in its bound form corresponds to its meaning when occurring as free forms, the resultants can be considered real compounds. For a comparison of the semantics of *over-* and *under-* see Plag (2003:222-223) and for *out-* see Adams (2001:75). The only example for type 14a is *overripe*, a much-cited example in literature which can be considered as a real particle-adjective compound (cf. Plag 2003:223).

Type-14b examples are just derived forms from particle-verb compounds (*overacting* from *overact* and *outstanding* from *outstand*) and are thus not genuine article-participial adjective examples.This seems to apply for most of the examples in type 14c as well: many of them have occurred in type 10 as particle-verb compounds. Questionable at least seems to be the example of *back-calculated* (*back-calculated concentrations, back-calculated pore water quality*). It only occurs in one text and has no compound verb equivalent (*calculated back*) in the whole corpus. There is just one instance of *backcalculation* which allows speculations that *backcalculated* is a result of a backformation process.

Type 15	verb + particle
	step-downs, trade-offs, pump-out, build-up, clean-up, punch-outs, push-outs
results in	compound nouns, compound adj.

Examples of this type are also only compounds according to their form of a combination of a verbal element and a particle. In reality they are compacted versions of verbal phrases which combine to form new nouns (*One method is to pump out the contents of each chamber. – assuming pump-out every four years,...* source: ENGICOR). Bauer describes them as follows: "The majority of words of this form are nominalizations of phrasal verbs, and it is arguable that these are not strictly compounds at all" (1983:208). And he and Huddleston (2002:1654) find that "this is still a productive process", a statement that can be confirmed by a number of corpus findings which are either nouns (*pump-out, bake-out, build-up*), or both, used as nouns and in attributive position (*scale-up, drop-out*).

Multiword compounds

Up to now, only two-word compounds have been examined. Compounds are considered binary structures (cf. Plag 2003:133), and multiword-compounds consisting of more than two words comply to this rule as well: for *waste rock dumps* there are not the three words put together in one process but *dumps* that is added to *waste rock.*

At least with longer compounds consisting of a number of nouns, this rule can be applied, although it is sometimes hard to tell which compounds are added in which order. For *household hazardous waste collection days*, I assume there was first the *hazardous waste* which was further specified by *household* to become *household hazardous waste*, where *collection* was added to describe the process of fetching the waste from the homes and as this happens at particular times, finally *days* was added to form *household hazardous waste collection days* (alternatively *collection days* might have existed before and was then added to *household hazardous waste*). But as the practicability in terms of ease of production and understanding of such words is limited, four-member words are rare and five-member words extremely infrequent compared to the vast amount of two-word compounds. For three-word compounds the following patterns could be established:

- n+(n+n) *tailings pore water*
- (n+n)+n *storm water runoff*
- adj+(n+n) *sustainable waste management*
- (adj+n)+n *hazardous waste generation*
- adj+(adj+n) *volatile organic compounds*

When we do not look only at nouns and adjectives forming multiword compounds, we come to the interesting phenomenon of phrasal compounds and their environmental engineering English specificity.

Phrasal Compounds

Phrasal compounds are what I would consider to be compounds whose noun elements are linked by a preposition. According to Fischer (1998:44) a "lexical phrase designates a word-formation type which occupies an interim position between the word and the syntactic phrase".

They describe domain-specific concepts and are used to express a complex idea in a relatively short way. Phrasal compounds expressing a ratio between two substances or parameters is the most frequently used: i.e. *air-to-fuel ratio, ash-to-plastic ratio, area-to-volume ratio, mass-to-charge ratio*. If described differently, longer phrases would be needed: *pay-for-performance basis* – paid according to the quality of the results or *resistance-to-pressure test* – a test which measures the resistance to pressure. Other examples of phrasal compounds are:

- *well-within-well solutions*
- *sand-in-sand filters*
- *end-of-pipe control*
- *glass-to-fines processors*
- *fiber-to-fiber contact*
- *behaviour-over-time graphs*
- *method-of-drive recommendations*

The structure of their first element usually is noun+preposition+noun used attributively before the head noun. The following example of *waste-to-energy* deserves special mention because of its varied occurrences it can be considered as series-forming whereas the above examples look like rather infrequent on-the spur creations.

The "*waste-to-energy*" type manifests itself in the following examples produced by analogy processes:

- *waste-to-energy processes/facilities/options*
- *landfill gas-to-energy project*
- *landfill-gas-to energy industry growth/developer/power plants*
- *LFG-to-energy recovery systems/projects/facilities*

- *tire(s)-to-energy plant/facilities,*

and will be active in parallel with the industry that evolves from this technology.

Phrasal compounds are a rare phenomenon but on grounds of their specificity I claim that they should be introduced to the learners of the specialist language because they contain implied shortened concepts (*landfill gas turned to energy*).

5.2.5 Combining Processes

Whereas other word-formation patterns occur in general language as well as in domain-specific language, combining processes can be said to predominantly be active in science and technology. In the different scientific domains the compact results of combining processes "convey a complex matter in a shortened form" (Fischer 1998:61) When specific knowledge from certain domains begins to enter common knowledge, at the same time, their domain-specific linguistic items gain access to the common language as in the case of *multimedia, biogas* and *eco-tourism.*

Combinatorial processes where combining forms like *geo-* and *–phyte* amalgamate with either a free base (i.e. *bioremediation*) or another combining form (i.e. *hydrophilic*) will be called here combining processes. Their place in the categorisation – between compounding and derivation – is deliberately chosen as it is this inbetween status - being in some respect close to compounding and in others to derivation - that makes combining an own word-formation pattern. This status is not uncontested and so we can find combining processes sometimes subsumed under derivational (cf. Quirk 1985:1545) sometimes under compounding processes (cf. Plag 2003:155 ff.) and sometimes being treated as a separate process after all (Bauer 1983:216). For an overview of different attempts to define combining processes see Warren (190:122 ff.)

The notion of a combining form first only extended to forms of neoclassical format originating from the Greek and Latin language. They can be added at the beginning of a free base or another combining form, hence we speak of initial combining forms (i.e. *microorganisms*), or to the end (i.e. *halophytes*) what we then call final combining forms. More recently, the term has been extended to products of secretion like *-scaping* (cf. Warren 1990:117 ff.) where a complex word like *landscaping* is shortened while only so much of the original meaning is kept "to provoke the right associations" and "to have the phonological characteristics of a morpheme suitable to be a word component" (Warren 1990:117/118). ENGICOR as a source for an instance of such a secretion process provides an example plus explanation for the form *xeriscaping* which is "landscaping with plants that use minimal water and generate minimal waste"; it is a word formed by an allomorph of the Greek-originated combining form *xero* (dry) and the shortened form *–scaping* (from *landscaping*). As this was the only attested example of this kind, the following investigation will be restricted to neoclassical combining forms.

Initial combining forms serve as modifiers, final combining forms as heads. None of them occur as a free form. They differ from affixes in that they are able to combine with other combining forms whereas affixes can only combine with bases. Another difference is the fact that "the meanings of combining forms are generally more specific than is typical of affixes" (Bauer/Huddleston 2002:1664) although some combining forms and meaning-related prefixes get very close (i.e. *mini-* and *micro-*). Consequently, distinguishing sometimes between an initial combining form and an affix can be difficult and leads to the fact that some items are classified as one or the other in different reference works (cf. Warren 1990:123).

Compared to the number and variety of initial combining forms, final combining forms are rarer in type and token frequency. In ENGICOR, they exclusively combine with initial combining forms. They form adjectives by the addition of the suffix *–ic* like *cryophilic* (low temperature loving), *hydrophobic* (water-fearing/water-repellent), *phreatophytic* (living on decaying organic matter), or *methanotrophic* (feeding on methane), and nouns like *halophytes* (plant adapted to live in saline conditions), *heterotrophes* (plants feeding on an external supply of energy), *batholith* (a particular rock mass), or *pneumatophore* (air-carrying plants). Except for *batholith*, all examples pertain to the field of plant morphology.

Initial forms often end in a vowel which is in most cases –o, in some –i, a further feature that distinguishes them from many prefixes. They are "open-set items which can be formed on the spur of the moment" (Warren 1990:125). In environmental engineering English, this creation on the 'spur of the moment' does not seem to be the predominant case. More important is the production of series of words referring to the same feature, such as *phytoabsorption, phytoextraction, phytoaccumulation, phytofiltration, phytomining, phytoremediation, phytodegradation, phytostabilization*, these manifold techniques involving the use of plants to remediate soil and groundwater.

Before I go into detail describing the determined initial combining forms, I want to have a short look at combinations with *–meter*. According to Warren (1990:118, 130) in a combining form that has a particularized meaning derived from a model word like *thermo-* or *barometer; -meter* stands for a measuring instrument. Rather than asking the question whether it qualifies as a combining form I want to concentrate on the fact that it combines with initial combining forms like *thermo-* and *hydro-* to result in *thermometer* and *hydrometer*. It has been quite productive according to the ENGICOR-derived list of different words produced by combining with *–meter*, i.e. *accelerometer, colorimeter, extensometer, olfactometer, piezometer, rotameter, viscometer, wattmeter*, etc.

The earliest forms of such kind are for instance *thermometer*, a Greek combining form + *-meter*. Later Latin-based first elements occurred with *–meter* (i.e. *calorimeter*) and also hybrid forms like *flowmeter* and *permeameter* where there is even no central –o or –i between the two elements. Using *–meter* to denote measuring instruments has been in fashion since the 17th century and I assume still might be because of items like *durometer, consolidometer, weightometer* and *turbidometer* which are not listed in the OED. Forms like *weighto-* and *consolido-* are evidence of the phenomenon that combining forms are an open-set class and can still be produced, albeit no longer on the basis of the classical but on the English language. We can find items originating in the combination of a genuine neoclassical combining element like *lysimeter* or *piezometer*, English-derived look-alikes like *diffractometer* (from the verb *diffract*) or *penetrometer* (from the noun *penetr(ation))* or instances where the English word has kept its original form in full (*flowmeter, wattmeter*) or in parts (*permeameter*).

5.2.5.1 Initial Combining Forms in ENGICOR

As already mentioned, initial combining forms outnumber final combining forms, thus are quite frequently employed in environmental engineering English. In this paragraph I aim to have a closer look at the meaning side of the combining forms under investigation as well as at the sort of second constituent they combine with and will try to establish which combining forms are more frequently used than others.

The first group of initial combining elements are ***micro-***, ***macro-*** and ***mega-***, which I grouped together on the grounds that they all indicate size; *micro-* standing for small, *macro-* for large and *mega-* for great.

Added to nouns, *micro-* indicates their small size like in *micro-sensors, microchips, micropores*, or *micro-cracks*. In most cases, *micro-* denotes the small-scale character of entities that have normal-scale equivalents as well, i.e. *microfauna, micro-climate, microenvironment*, or *micro-nutrients*. Some denote processes which help eliminate small particles like in *microfiltration* or *micropurging*. It is furthermore used to name units in physics and to a lesser extent chemistry (i.e. *microamp, microsecond, micromole*) and indicates the fact that some branches of sciences developed after the invention of the microscope (*microbiology, micromorphology*) or other technology that made applications involving minute size objects possible (*microelectronics*). Adjectives containing *micro-* are often derived from the corresponding nouns like *micro-porous* from *micro-pores* and *micro-climatic* from *micro-climate*. A word like *microfine* seems to indicate that the meaning of *micro-* is extended to a quantifying 'very' in these cases, its antonym *micro-rough* though implies that the rough surface is covered in particularly tiny rough bumps. The verb *microencapsulate* only reveals its meaning to the non-specialist when we understand that during the process of *microencapsulation* substances are enclosed in *microcapsules* from which the process noun and the verb is derived, as I assume *micropurge* is from *micropurging*.

Summarizing, we can say that *micro-* combines with nouns, adjectives and verbs to form new words. The type frequency of words starting with *micro-* is 58.

Macro- often forms opposites to combinations with *micro-; macro-pores, macroporous, macroencapuslation* and *macropollutants* are evidence of that phenomenon. Of the only 11 occurrences containing *macro-*, these four have their small-scale equivalents starting in *micro-*. Combined with *macro-* nouns indicate the (relatively) large size of the denoted objects as in *macropores, macrophytes* and *macroinvertebrates*. In some cases it designates a more comprehensive level of state as in *macro-level* or *macrodispersion*. Only two of the 11 instances are adjectives, all others are nouns.

The use of *mega-* is with only 9 types even more infrequent than *macro-*. In four of the nine cases it is used to form units of measurement (i.e. *megawatt, megajoule, megatonnes, megagrams*). In *megalandfill, megaproject, megacity* and *megamergers* it adds the meaning of very large to the nouns. *Megapopulated* is the only instance of adjectival use and here *mega-* somehow adds the notion of extremely populated.

The next three initial combining forms – ***mono-***, ***poly-*** and ***multi-*** - have in common that they indicate some sort of valency. *Mono-* stands for single or alone, *poly-* and *multi-* for many. Another feature that only *mono-* and *poly-* have in common is that – together with the prefixes *di-* and *tri-* - they can be part of a series-forming nomenclature of chemical compounds like *monomer, dimer, trimer, and polymer* or *monomethyl, dimethyl, trimethyl~, and polymethyl~*. Other non-series forming examples are for instance *monosodium, monoxide, polybutadiene, polyester*, and *polyvinyl*. *Mono-* furthermore combines with other nouns to form new language items like *monolayer, mono-fracture, monofill, monoculture*, all implying that the object in question comprises just one specific compound. Similarly, adjectives starting with *mono-* describe objects that possess only one single item of the named entity: *mononuclear, monoatomic, monoprotic* (possessing one single proton). Other adjectives might be derived from nouns: *monochromatic* and *monomeric* for instance. *Mononucleosis* is an example of a derived noun from *mononuclear.*

Apart from the already mentioned chemical names, ENGICOR does not yield many instances of *poly-* combining with nouns. The only ones are *polycoat, poly-phase*, and *polydispersity*.

In parallel with *mono-*, we can also find adjectives of the kind *polynuclear, polyprotic, polyaromatic*. Furthermore, we have chemical nouns on whose basis other nouns, verbs and participial adjectives derive like *polymerize, polymerization* and *polychlorinated*.

Examples for adjectives with initial *multi-* are *multidirectional, multidimensional, multiphasic, multicellular*, and *multivariate*, all indicating that the modified nouns dispose of many of the mentioned items (many directions, many dimensions, etc.). Participial adjectives also combine with *multi-* as in *multifaceted, multi-layered, multi-lobed* or *multi-tiered*.

Combined with nouns, *multi-* adds the sense of many or multiple as in *multiaxle, multibin, multiplume, multi-unit,* and *multi-level*. Many of the resulting *multi-*+noun combinations are used attributively (*multi-barrier concept, multifamily homes, multilevel samplers, multi-pollutant control*, etc), which further distinguishes *multi-* from *mono-* or *poly-*.

Moreover, unlike *mono-* and *poly-*, *multi-* is not of Greek but of Latin origin.

Most items with *mono-* and *poly-* combine with words of foreign origin (Greek and Latin) and only a few of English origin (*monorail, monolayer*). In contrast to that the majority of words combining with *multi-* are of English origin (compare *monochrome* and *multi-colors*).

The next initial combining form under investigation is ***auto-***. Because of its meaning – self, by oneself, independently, self-acting or automatic – it does not fit any of the other groups of combining forms, and is therefore here treated separately. *Auto-* combines with words and final combining forms (i.e. *autodialer* and *autotroph*), with words of foreign and of English origin (*autooxidation* and *autosampler*) and in terms of word class with nouns, verbs, adjectives and adverbs (i.e. *autodecomposition, autodecompose, autoimmune, automatically*).

In nouns like *autodecomposition, auto-degradation, auto-oxidation*, the meaning component 'self' is prominent whereas in *auto-tarping* or *autoclaving auto-* stands more for *automatic*.

The automatic-component is also be found in self-acting technological equipment like *autodialer* and *autosampler*. Examples of adjectives indicating that something is acting itself is *autoimmune, autotrophic* and *autothermal*. The verbs are either derived from pre-existing nouns (*autoclave v.* from *autoclave n.)* or like *auto-tie,* which must have come into being independently of any noun.

Under the next group of initial combining forms I subsume the following combining forms: ***bio-, phyto-, geo-, eco-, photo-, electro-, thermo-, hydro-***, and ***rhizo-***. They all have in common that their meanings denote an entity in the field of science that has to do with natural phenomena (i.e. light, heat, water, electricity) or related objects that are central to the domain of environmental engineering (plant, root, and earth). It is a way of grouping according to their meaning load that allows *eco-* to be included in it as it is no combining form but the clipped form of *ecological/ecology*. I chose to include *eco-* here also because with *bio-* and *hydro-* it is not always straightforward to say whether a particular instance originated from a combining process or the clipping of *biological/biology* or *water* and in parallel with a combination with another stem, which we will see later.

All combining forms discussed here are of Greek origin; *ecological/ecology* which is shortened to *eco-* is also of the same origin. Furthermore, except for *rhizo-* they occur all together with words of English origin as well as of foreign origin although the pattern combining form + word of foreign origin is much more frequent in general.

There are very few cases where an initial combining form in this group combines with a final combining group (i.e. *thermophiles, hydrophilic, photography, geology,* etc.). Words like *biology, ecology* and *hydrology* were eliminated from the count as they might have served as a source word for a clipping process.

It is above all new nouns that are produced by the addition of the combining forms of that group. In these cases the combining form modifies the nominal head and produces hyponyms of already existing terms (*degradation – biodegradation, toxicity – phytotoxicity, efficiency – eco-efficiency*). On the basis of some words co-hyponyms are produced: *biodegradation, phytodegredation, photodegradation, rhizodegradation.*

ENGICOR yields only 9 verbal instances involving an initial combining form of this group (though it is not claimed that the other combining forms produce more, verbal forms are rare overall). Adjectives are more frequent and more varied than verbs. They usually have a modifier-head structure but some like *electro-chemical* or *thermo-mechanical* are dvandva adjectives were the two constituents are semantic coordinates.

There are a few instances of adverbs containing one of the combining forms which derived from their corresponding adjectives.

Modifying *bio-, phyto-, photo-, electro- hydro-* and *rhizo-* combine with nouns denoting processes in environmental engineering like *bioremediation, phytomining, photo-oxidation, electroplating, hydro-pulping* and *rhizofiltration.* As in the semantic classification of noun compounds (5.2.4.1.3) combining processes can also yield nouns which function as process indicators or process parameters like *bioavailability, bioconcentration, phytotoxicity, geosmin, electroneutrality* and *hydrophobicity.* We furthermore have process agents and products of processes like *biochemicals, phytomass, eco-sand, thermoplastics* and *hydrocompost* and process equipment and technical devices like *biofilters, geopipes, photocells, thermocouples* and *hydroseparators. Bioplume, ecosystems, thermo-* and *rhizosphere* are generally said locations where chemical, physical or biological processes take place.

To continue beyond the constructed schemata semantic classification used in the compounding section, some initial combining forms of this group form words denoting sciences or branches of science like *geomorphology, photochemistry, electrokinetics, thermodynamics* and *hydrogeology*. There are a few instances of designations of professions like *geologist* and *geomorphologist*, and other living agents like *thermophiles* and *photoautotrophs*.

It might be due to the semantics of the clipped form *eco-*, that some words resulting from combining with it do not follow the pattern in the grid and leave *ecofees, ecotax, ecototourism* and *ecocapitalism* a bit aside.

I assume that most verbs are derived from their corresponding nouns like *bioaccumulate, biodegrade, phytoextract, phytoremediate, photodissociate* or *electrowin*. Their type and token frequency is very low.

When it comes to forming adjectives with combining forms, the combination with normal adjectives is particularly productive (i.e. *bioavailable, phytotoxic, geothermal, eco-efficient, photodegradable, thermoreactive, electroactive, hydrogeologic, rhizospheric,* etc). In contrast to this variety, participial adjectives in *–ed* and *–ing* are quite rare (*biotransformed, bioaccumulating, phytostabilizing, photoluminiscenting, thermoforming, thermosetting*).

Phyto- and *rhizo-* seem in particular to have widened their domain-dependent spectrum as their entries in the OED indicate that they are chiefly used to produce botanical terms. The evidence from the corpus shows that they clearly belong to the domain of environmental engineering producing new words in this field (*phytodegradation, phytoextraction, phytoremediation, phytovolatilization, rhizodegradation, rhizofiltration,* etc.). It has already been mentioned above, that *eco-* in particular but also *bio- and hydro-* to some extent cannot be called a combining form proper, but are results of clipping processes. Corpus evidence for that phenomenon comes in the form of doublets like:

• biological degradation (11x in ENGICOR)	• biodegradation (165x)
• biologically degrade (1x)	• biodegrade (36x)
• biologically degradable (1x)	• biodegradable (135x)
• biological barrier(s) (4x)	• biobarrier(s) (24x)
• biological clogging (3x)	• bioclogging (2x)
• biologically active (21x)	• bioactive (5x)
• biological activity (28x)	• bioactivity (3x)

For items like *biodegradation*, where the frequency of the shorter form is so high, it cannot be assumed that they developed from the long form *biological degradation*. Others like *bioclogging* and *bioactive/bioactivity* are more likely candidates for such a route. In any case, what these numbers show is that the meaning of the combining form *bio-* stemming from Greek bios (meaning life) is so close to the meaning of the assumed clipped form of *biological* or *biology* that users of language do not make the difference and therefore use it as if there is no difference. This might well contribute to the high productivity of *bio-*.

With other items again like *biocide* or *biosphere*, the possibility of an existence of a long form can be ruled out on grounds of the structure and morphological origin (**biological cide, *biological sphere*).

For *hydro-* only one case of it being the shortened form of *hydrogen* could be established: *hydrocyanic*, derived from *hydrogen cyanide*.

5.2.5.2 Productivity of Initial Combining Forms

Having looked at the meaning side and at the combinatorial characteristics of the combining forms I will now try to answer the question which forms have been or still are more productive than others in the field of environmental engineering English.

The rationale behind this question is the idea again that students should be exposed to the most salient processes and develop a certain awareness where there is ongoing language change expected.

Morphological productivity is a much disputed topic (cf. Bauer 1983:1-6, van Marle 1992), although a general definition as to what it implies is easy to figure out: it can be described as the ability of a word-formation process to coin new words. The underlying assumption there is "that people can say things which they have neither said or heard before because they know (albeit not explicitly) the rules for the combination of appropriate elements into sentences and because they are able to assign meanings to those elements" (Bauer 2001:1-2).

As most modern attempts bent on productivity tend to be a scalar rather than a yes-or-no phenomenon (cf. Bauer 2001, Plag 1999); the crucial question is how it can be determined. Scholars who tried to operationalise a procedure to measure productivity have either (1) used a dictionary-approach like Caso, Algeo and Bauer mentioned in the introductory categorisation chapter where they just counted type-frequency of the different word-formation patterns, or (2) or a corpus like Baayen and Renouf (1996) who established various quantitative measures based for instance on hapax- and token frequency, and (3) a combination of the two like Plag (1999).

Following the ideas of the various frequency counts, I want to go through the different measures step by step and evaluate which one is most suitable for giving an indication which items are worth concentrating on in the ESP class.

At the most straightforward level using a corpus we have token frequency (i.e. how often words with for instance the combining form *phyto-* occur in the corpus) and type frequency (i.e. how many different words get formed with *phyto-*) (compare Table 19). Both play a role in the notion of text coverage, that is how often learners meet *phyto-* as combining form in a text and how often they meet a new (new in the sense of the text) word in the text. In terms of a high token frequency students of environmental engineering English quite often encounter *bio-*, *geo-*, *micro*, *auto-* and *phyto-* in the running text of the corpus (although we can of course not assume even distribution). Every 572nd word would be one starting with *bio-*, every 614th with *phyto-*. Looking at the type frequency, *bio-* and *micro-* are also very successful candidates – producing various different words beginning with them. In contrast, *geo-* and *auto-* and even more so *phyto-* do not produce such a variety of different words; so in terms of type frequency they are less productive. *Multi-* and *photo-* are examples of the opposite way: they display fewer tokens but rank higher up in type frequency, so we can expect them to combine with many different language items.

Tab. 17: type and token frequency of combining forms

combining form	**number of tokens in ENGICOR**	**number of running words in ENGICOR/ number of tokens**	**combining form**	**number of types in ENGICOR**	**number of running words in ENGICOR/ number of types**
bio-	3,493	572	bio-	115	17,391
geo-	1198	1,669	multi-	66	30,303
micro-	714	2,801	micro-	58	34,482
auto-	708	2,824	photo-	39	51,282
phyto-	614	3,257	geo-	37	54,054
electro-	475	4,210	auto-	31	64,516
thermo-	423	4,728	electro-	31	64,516
eco-	306	6,535	mono-	27	74,074
multi-	265	7,547	phyto-	25	80,000
mono-	255	7,843	thermo-	23	86,956
photo-	180	11,111	eco-	14	142,857
rhizo-	116	17,241	macro-	11	181,818
macro-	33	60,606	mega-	9	222,222
mega-	29	68,965	rhizo-	4	500,000

The approach of using type and token frequency to determine productivity was used in the Longman Grammar of Spoken and Written English (Biber et al. 1999). Here the total number for instance of noun lexemes with a certain word-formation element are assessed and they claim that "it is useful to distinguish between common and rare nouns, in that rare nouns are more likely to reflect either recent coinages gaining currency or an older coinage becoming obsolete" (1999:323).

So in terms of frequency, language teachers and learners should concentrate on the words that rank high up in token and/or in type frequency. This measure, however, does not answer the question which combining forms are particularly productive and of which we can expect more language change in the near future of environmental engineering English.

Baayen and Renouf (1996:74) claim that "if a word-formation pattern is unproductive, no rule is available for the perception and production of novel forms. All existing forms will depend on storage in the mental lexicon. Thus, unproductive morphological categories will be characterized by a preponderance of high-frequency types, by low numbers of low-frequency types, and by very few, if any, hapax legomena, especially as the size of the corpus (...) increases." This view assigns crucial importance to the role of hapaxes – words that occur only once in a corpus - saying that they serve as an indicator of the productivity of the word-formation pattern in question, for it is assumed that it is amongst the hapaxes that there are the most neologisms. Table 20 provides the number of hapaxes of the different combining forms in ENGICOR. Again it is *bio-* which displays the most be followed by *multi-*, *micro-*, *photo-* and *auto-*. *Rhizo-*, *mega-*, *macro-* and *eco-* do not yield many hapaxes and would therefore considered to possess a lower degree of productivity because of this measure. Baayen developed this quantitative corpus-based approach further when he introduced different formulas to measure different aspects of productivity (cf. Baayen 1991,1992). I will here use his formula for "productivity in the narrow sense" to find out whether it will provide further insight into the question which combining forms are more productively used than others.

Productivity in the narrow sense is the quotient of the number of hapaxes of a given affix divided by the number of tokens with this given affix and should thus serve as an indicator for morphological productivity. Looking at the results for the initial combining forms, *mega-* shot to the top of the ranking, followed by *multi-* and *macro-*, *bio-* lost much ground and forms like *electro-* or *mono-* stayed almost constant in the middle. With forms like *mega-* and *macro-* no secure claims can be made, as their data is too scarce. It is doubtful to take them into account, as their overall frequency does not allow clear assumptions. *Multi-*, however, shows what it needs to rank high up in productivity in the narrow sense: a large number of hapax legomena and a low number of tokens. Although the number of hapaxes of *multi-* and *bio-* are almost identical, the fact that *bio-* has a very high token frequency results in *bio-* being at the bottom of the list, indicating that it is not (very) productive.

Tab. 18: number of hapaxes/productivity in the narrow sense

combining form	**number of hapaxes**	**number of running words in ENGICOR/number of hapaxes**	**combining form**	**productivity in the narrow sense**
bio-	36	55,555	mega-	0.1379
multi-	34	58,823	multi-	0.1283
micro-	28	71,428	macro-	0.1212
photo-	17	117,647	photo-	0.0944
auto-	15	133,333	mono-	0.0431
phyto-	12	166,666	micro-	0.0392
electro-	12	166,666	electro-	0.0252
mono-	11	181,818	eco-	0.0288
geo-	9	222,222	auto-	0.0211
thermo-	8	250,000	phyto-	0.0195
eco-	7	285,714	thermo-	0.0189
macro-	4	500,000	bio-	0.0103
mega-	4	500,000	geo-	0.0075
rhizo-	0		rhizo-	

Such purely quantitative approaches to measuring (indicators of) morphological productivity are much disputed (cf. van Marle 1992:151ff., Bauer 2001: 149ff.). Examples - as in the present case *bio-* - run counter to the idea that productivity can validly be measured by means of statistics applied to corpus investigation. Van Marle expresses his criticism as follows: "Once a word is coined, the frequency in use of that word, it seems to me, is more or less irrelevant to the degree of productivity of that rule" (1992:156). Bauer adds that it is a "mistake to equate frequency and productivity" (2001:20).

Supporting these doubts on the grounds of the example *bio-*, it can be asked whether a purely quantitative approach can provide the sought information on which items to particularly focus on in class.

For this reason, the corpus-based quantitative approach is left aside for the moment and I will turn to a qualitative, dictionary-based investigation using the OED (Oxford English Dictionary).

All words containing one of the above-mentioned combining forms were looked for in the OED and it was noted whether it is listed there or not and if so, whether the first mention was before or after 1900. The following table displays the number of items of the concerned combining form found in the OED after 1900 and the number of items not listed in the OED:

Tab. 19: occurrence of combining forms in OED

combining form	listed after 1900	combining form	not listed
multi-	31	bio-	84
photo-	24	multi-	29
micro-	22	phyto-	25
bio-	21	micro-	21
electro-	11	geo-	14
auto-	10	auto-	11
thermo-	9	eco-	8
mono-	7	mono-	7
macro-	4	macro-	6
mega-	3	electro-	5
geo-	2	mega-	5
eco-	2	thermo-	4
rhizo-	1	rhizo-	3
phyto-	0	photo-	1

The OED-search provided a differentiated picture. It shows that combining forms that can be related to older and more established technologies like *photo-* and *electro-* rank up higher in the column "listed after 1900" but low in the column "not listed"; consequently it can be assumed that they are not very often used to coin new words. In contrast to that, none of the items containing *phyto-* is listed after 1900 but 25 items of *phyto-* were not found in the OED at all. What about our watershed-case: *bio-*? It has a considerable number of forms listed after 1900, but with 84 occurrences which do not occur in the OED it is absolutely top of that list.

Criticism on the use of dictionaries for representative sampling of morphological patterns is legend (cf. Bauer 1992, Baayen/Renouf 1996:69, Plag 1999:97-98) because the choice of words to include is arbitrary, items en vogue might be included to a large part, others simply overlooked. However, the OED as the largest and most comprehensive historical dictionary has - particularly in the most recent additions from 1993 and 1997 - included entries that cover "subjects that have recently gained prominence" as well as "main technical vocabulary" (source: Preface to the 2nd Edition). In the light of the background of still very limited specialist dictionaries for environmental engineering English, and after intensive work with the OED, I am convinced that it makes a superb source of specialist vocabulary information. The whole gamut of entries included is vast. Still, the question remains whether the words not listed in the OED have really been neologisms at some stage or whether they are just rare words that have not found entry into the OED.

In order to find this out, Google was searched for the number of hits of ENGICOR-words with the combining form *bio-*. The outcome clearly supports the claim that the words not listed in the OED are more recent formations. The following table displays all words from the corpus containing *bio-* together with their number of Google hits. Those items which are not listed in the OED are in bold, those which are listed after 1900 are in italics and those occurring in the OED before 1900 are left normal. There is the obvious tendency (1) of the items not listed to have fewer hits, (2) the ones listed after 1900 start occurring in the middle and cluster towards the end, and (3) those being established and mentioned in the OED before 1900 cluster at the very end, displaying a huge number of hits. Moreover, even a non-specialist can tell that these words belong to technologies which have recently been developed in the field of environmental engineering.

Tab. 20: occurrence of combining forms on the internet

items starting with bio-	**Google-hits**	**items starting with bio-**	**Google-hits**	**items starting with bio-**	**Google-hits**
bioaerosolization/sation	9	**bio-degradables**	7860	**bioactivity**	209,000
bioassimilable	65	**bio-accumulating**	10,500	**bioaccumulation**	225,000
biowalls	79	**biodegrading**	11,100	**bioavailable**	249,000
bioaugmenting	97	**bioswale(s)**	11,100	**biocontrol**	268,000
biofilming	117	**biosurfactant**	11,800	*biocide(s)*	317,000
biodosimetric	280	**biodegradative**	11,900	**biotransformation(s)**	364,200
bioclogging	335	**biotransformed**	12,900	*biogas*	375,000
biorefractory	349	**bio-oxidation**	14,380	*biogeochemistry*	386,000
bioflocculation	364	**biostimulation**	17,500	biochemically	437,000
biorecalcitrant	405	**bioretention**	19,300	**biofilm(s)**	471,000
biocatalyze/ise	412	**bio(-)venting**	19,600	*biogenic*	492,000
bio-oxidizer(s)	495	**biotoxin**	22,300	**bioenergy**	515,000
bioattenuation	506	**biotreatment**	23,000	*biomolecules*	526,000
biogrowth	510	**bioaugmentation**	25,700	**biosynthetic**	544,000
biogasification	554	**biomat**	30,500	**bioactive**	547,000
bio-uptake	620	**biorefinery/ies**	31,300	*biogeochemical*	569,000
bioslurry	757	**bioacatalyst**	33,300	**biochemicals**	710,000
bioremedial	767	**bio(-)waste**	34,000	*bioavailability*	713,000
bio-approaches	775	**biomagnification**	35,900	**bioavailable**	987,000
biorestoration	800	**biocatalytic**	41,500	*biodiesel*	1,050,000
bioremediated	834	*biocidal*	47,000	*bioassay(s)*	1,070,000
bioremediating	890	**biofiltration**	53,100	*biofuels*	1,472,000
biocarrier(s)	1140	**bioconcentration**	57,800	*biomechanics*	1,560,000
bio(-)towers	1271	**bio(-)fouling**	59,400	**biophysical**	1,590,000
biostimulating	1460	**bioconversion**	60,300	*biosynthesis*	2,170,000
biostabilization	3650	**bioaerosol(s)**	65,500	biosphere	2,270,000
biomining	4030	**bioaccumulate(s)**	73,900	*bioengineering*	2,650,000
bioremediate	4240	**bioindustry/ies**	83,800	biologically	3,840,000
biopiles	4274	**bioaccumulative**	94,400	biologist	4,970,000

biobarrier(s)	4710	**bio-catalysis**	101,000	biochemical	9,530,000
biomagnifies/d	4753	*biodegradability*	117,000	*biodiversity*	13,000,000
bioreactive	5140	**biodegrade/s/d**	120,100	biochemistry/ies	24,208,690
bio(-)sparging	5680	**biosource**	124,000	*biomedical*	44,800,000
biophase	6010	**bio-based**	152,000	*biotechnology/ies*	46,611,000
bioreduction	7390	**bioproducts**	174,000	biology	111,000,000
bioreaction	7650	**biomonitoring**	178,000		

5.2.5.3 Conclusions on Combining Forms

The above investigation into the productivity of some combining forms showed that a combination of quantitative and qualitative methods can provide a better insight into which combining forms are actively producing new words at the moment than ones which would have a sheer quantitative character.

As it cannot be excluded that the reason for the failure of the quantitative measures is due to the small size of the corpus, it can only be stated here that the usefulness of the productivity in the narrow sense is very limited when applied to such a small corpus. However, the findings allow room for doubts as to whether statistical measures are applicable in this field as there is a great number of constraints on the phonological, morphological, syntactic, semantic and pragmatic field (cf. Bauer 2001:126ff.) which hinder the production of possible words and which can in my opinion not be equated to numbers. Other authors like Plag, having compared corpus and dictionary results, come to a similar conclusion: "Although the quantitative results are useful and important, the methodological problems involved with both types of data-bases are enormous, and no single measure should be overinterpreted in its estimation of the productivity of a process" (1999:117).

By using two dictionary-based parameters – the number of words listed after 1900 and the words not listed in the OED – I tried to minimise the source of bias brought in by the fact that lexicographers might have overlooked one phenomenon and overstressed another.

In sum, it must be said that the results from the type and token counts can serve as an indicator of which language items to focus on in the ESP-class based on the fact that items with a high token frequency have a high text coverage, i.e. will be often encountered by the students. Also type frequency has its own importance in that words with a high type frequency are also of particular interest as they yield a high number of different types with the combining form in question, which is not necessarily the case with a high token frequency.

It is furthermore beyond doubt that it is amongst the hapaxes and according to the corpus size particularly among low frequency types where we find the most neologisms. Generally assumed as a means of obvious ranking, the resulting quotient of the number of hapaxes of a given affix and the number of tokens with the affix – the productivity in the narrow sense – yielded a ranking that was in parts very different from the qualitative analysis of the concerned items in the OED. Consequently, it can be stated that the numbers of hapaxes can be of certain interest but should not be overestimated, particularly for items with a low amount of data available. Productivity in the narrow sense seems to particularly fail those items which are very frequently used in the text so that despite a very high number of hapaxes, the high number of tokens brings down their productivity in the narrow sense to a very unrealistic value.

This was the point when a qualitative view was needed to make better-informed claims about morphological productivity. A combined approach, first locating candidates for neologisms with the help of the OED and then checking the likelihood of any of these terms to really be a word that has recently come into use by establishing their Google hits revealed different aspects of the productivity of particular language items. It describes a picture where every combining form must be looked at individually, as constraints on the different linguistic levels like phonology, morphology, etc. apply to each item in a very specific way which makes approaches to a straightforward number-based ranking on the scale of productivity a very challenging if not unlikely task.

Consequently, it might not be the application of the statistics involved that must be sophisticated but the subject of investigation: the corpus. Bauer describes such an "alternative view" as follows: "Another way to view productivity is in terms of rate of additions. The more productive a morphological process is, the more coinages that occur created by that morphological process in a given time period." (2001:156). So the alternative can be a large monitor corpus which can be examined in time slices and language items displaying certain word-formation patterns classified in the order of their occurrences. This approach is not new; it has been applied in the AVIATOR project (cf. Bauer/Renouf 1996). Another example of this method is Hänchen's (2002) investigation into the French marketing language where she searched through different years of a specialist journal for newly occurring language items. In order to make relatively secure claims on the exact data of entry of a neologism into a (specific) language, the amount of language data, i.e. the corpus, must be very large. ENGICOR, as its texts were only put together on grounds of the topics they represent and not according to years, cannot serve this aim. Nevertheless, it must be noted that it is a suitable source for the examples of word-formation patterns.

In the light of the short-comings of the corpus as well as the methodology at hand it was not possible to produce a linear ranking of the combining forms to illustrate their importance. What is however possible to say on the grounds of the study is that particular emphasis should be given to the combining forms like *bio-*, *multi-*, *micro-* and *phyto-* and that we can expect further neologisms from *thermo-*, *auto-* and *electro-* as well. *Macro-* and *mega-* might yield one or the other occasional new word but have no high overall productivity. In class, however, combining forms like *rhizo-* which only apply to a small number of possible bases should not be overlooked because they are quite specific and further development in technology might produce more items including *phyto-*.

Generally, what becomes clear from the combining forms is that as well as the linguistic conditions, extra-linguistic factors like the advancement of science and technology and for our example in environmental engineering bring along the new words. As environmental engineering is a booming branch we can expect more *bio-*, *phyto-* and *rhizo-* words to follow.

5.2.6 Derivational Processes

Derivational processes subsume all word-formation processes where an affix is added to a base word (mostly free, sometimes bound bases). The phenomenon of derivation itself is not environmental engineering English-specific as affixation is also a common means of word-formation in General English, but we will see that it is the extent to which certain affixes are used that is domain-specific.

The task of prefixes is to modify the meaning of the word to which it is added (compare *application* and *misapplication*). Suffixes have in general less semantic meaning but usually change the word class of the original word (compare *recycle* (v.) to *recyclable* (adj.). The semantic load some suffixes add is more specific than others (compare the *–y/-ey* in *clayey, dusty, slimy* – carrying the meaning of having the qualities of or being full of - with the much weaker meaning of *–ment* like in *containment, requirement* or *treatment*).

Affixes can be attached to nouns, adjectives, verbs (and adverbs). There are only a few affixes that can attach to all four word classes (like *pre-* in *pre-operational, prefiltration, pretreat, pre-operationally*).

The first part of this chapter on derivation aims at giving an overview of the different affixes found in ENGICOR, the bases they attach to and their semantics;, the second looks at some more specific issues relating to affixes which are particularly productive in environmental engineering English.

In the following descriptive part of the attested affix forms, the order of ranking is according to the input word class of the base to which the affix attaches. Then the semantic load of the affixes is discussed. I will begin with the prefixes.

5.2.6.1 Prefixes

5.2.6.1.1 Prefixes with Negative and/or Reversative Sense

non- 1497x in ENGICOR

non- + adjective > adjectives: (including participial adjectives)

- *nonaerobic*
- *non(-)hazardous*
- *non-environmental*
- *non-permeable*
- *nonporous*
- *non-reinforced*
- *nonnitrifying*

non- + noun > nouns:

- *nonapplication*
- *non-biosolids*
- *non-container*
- *non-contaminant*
- *non-curbside*
- *non-water*

There are only two instances of *nondestructively*, that is of non- + an adverb. *Non-* added to adjectives and nouns forms are quite established and therefore are frequently in ENGICOR occurring language items like *non-aqueous, nondurable, non-hazardous, non-point, non-profit* and very rare words like *non-aerosol, non-ash, non-avoided, non-channeling, non-clog, nonfoamed,* or *non-ICR*. *Non-* + noun derivatives are mostly used in attributive position (i.e. *non-lead fillers, nonmethane organic compounds*, etc.).

Non- is a negative prefix with the meaning 'not': *nonhazardous* > something which is not hazardous, *non-biolids* > something that does not belong to biosolids.

un- 2690x in ENGICOR

un- + adjective> adjectives: (including participial adjectives)

- *unavoidable*
- *unconfined*
- *undesirable*
- *undissolved*
- *unchanging*
- *unstable*

un- + noun > noun:

- *unavailability*
- *unemployment*
- *unrecyclables*
- *unsustainability*

un- + verb > verbs:

- *uncover*
- *unload*
- *unmask*

un- + adverb > adverbs:

- *undoubtedly*
- *unexpectedly*
- *unfortunately*
- *unnaturally*

Un- is used to express opposites to positive words like *avoidable* > *unavoidable*, *employment* > *unemployment*, *fortunately* > *unfortunately*. With verbs it conveys the meaning of a reversed action: *cover* > *uncover, load* > *unload*. In the majority of instances found in the corpus, *un-* combines with adjectives and participial adjectives.

The use of *un-* with present participles only extends to very old and lexicalised examples like *unwilling*, or *unwitting*, which cannot be considered domain-specific. erbs and nouns beginning with *un-* are extremely rare. There are a few adverbs with *un-* which, however, developed from the already negative adjective (*unnatural* > *unnaturally*).

in-/im- 2034/531x in ENGICOR

in-/im- + adjective > adjectives: (including participial adjectives)

- *immissible*
- *inorganic*
- *inaccessible*
- *inactivating*
- *incomputable*
- *inconclusive*

in-/im- + noun > nouns:

- *imbalance*
- *impurity*
- *inability*
- *insolubility*
- *intolerance*
- *inactivation*

in-/im- + adverb   > adverbs:

- *inproperly*
- *inadequately*
- *incompletely*

Words prefixed with *in-/-im* are predominantly adjectives, some nouns - most of them ending in *–ity* - and some adverbs from the belonging adjectives. If we do not count verbs which originate in the French language like *immobilize*, no verbs are formed with *in-/im-*. All instances are examples for the negative use of the prefix designating the opposites of the positive original words (i.e. *organic* > *inorganic, balance* > *imbalance, properly* > *improperly*). Most of the nouns beginning with *in-/im-* seem to be derived from their adjectives. Within our examples just imbalance does not have such an adjective as starting point.

dis-

For *dis-* I abstained from giving the frequencies of derivations starting with it as it is in many cases impossible to tell which ones are genuine English words and which entered the English language via French (i.e. *dismount*).

dis- + adjective   > adjectives: (including participial adjectives)

- *disqualifying*
- *dissatisfied*
- *disabled*
- *disagreeable*

dis- + nouns   > nouns:

- *discoloration*
- *discontinuation*
- *disinfectant*
- *displacement*
- *dissatisfaction*

dis- + verb   > verbs:

- *discolor*
- *disinfect*
- *displace*
- *disregard*

All adjectives starting with *dis-* are the opposites of their positive counterparts like *disqualifying* > *qualifying, disagreeable* > *agreeable*, implying the meaning of the negative 'not' in them. When attached to nouns or verbs *dis-* has reversative meaning, implying an undoing of a process, i.e. *coloration* > *discoloration, place* > *displace*. There is only one instance of an adverb with *dis-* (i.e. *disproportionately*) derived from the adjective *disproportionate*.

de-

What was said for *dis-* in terms of difficulties in deciding which words to include in a frequency count and which not, applies to *de-* as well. Some words are clearly affixed in the English language (i.e. *debagged*), others came via the French language of which some are even still transparent in English (cf. *decompose*).

de- + adjective> adjectives: (participial adjectives)

- *de(-)aired*
- *de-asphalting*
- *dehumidified*
- *deinking*
- *deprotonated*

de- + noun > nouns:

- *deactivation*
- *deanimation*
- *debagger*
- *defroster*
- *demanufacturer*
- *dewaterability*

de- + verb > verbs:

- dealuminate
- dechlorinate
- decolorize
- delist
- denitrify

De- can combine with adjectives (preferably participial adjectives), nouns and verbs, all implying the sense of removing or reversing which designate process-oriented activities typical for environmental engineering. There are no examples of normal adjectives attested; the existence of *dewaterable* can only be assumed on the grounds of the existence of *dewaterability*. The other found nouns are either words for processes or for persons or instruments involved in reversing or removing processes. Most verbs are used to describe "undoing" processes where the chemicals which contributed to the words are involved. Adverbs with *de-* were not established.

a-/an-

For the suffix *a-*, it was also felt that the effort of manually sorting all words beginning with this suffix from the more than 250,000 beginning with an 'a' was not justified in terms of the small outcome.

Attested examples are nevertheless *abiotic, anaerobic, aseptic* and *atypical*, all de-adjectival forms denoting the opposite of their positive forms.

5.2.6.1.2 Prefixes with Temporal and Spatial Sense

re-

Here again the line between genuine English formations and the French and Latin influence is difficult to draw, hence I will not give the frequencies.

re- + adjective > adjectives: (including participial adjectives)

- *recleaning*
- *recoverable*
- *recyclable*
- *refillable*
- *renewable*
- *reinforced*

re- + noun > nouns:

- *reactivation*
- *reconstruction*
- *recontouring*
- *recyclability*
- *recycling*
- *redevelopment*

re- + verb> verbs:

- *reaffirm*
- *rebuild*
- *recirculate*
- *redesign*
- *relocate*
- *reproduce*

Re- is a prefix whose activity is centred around verb-forming activities. All of the above examples for nouns and adjectives starting with *re-* are rather evidence for the productivity of the respective suffixes than for *re-*. The *recycle*-family is a good example to illustrate this phenomenon (*recycle > recycling > recyclable > recyclability). Recoverable* is not a sign of *re-* being active to form derivatives with adjectives but for *–able.* The general meaning of *re-* is 'again' or 'back' which can be seen in verbs of action like *rebuild* or *reproduce* again concentrating on actions and processes in environmental engineering.

pre-

pre- + adjective > adjectives: (including participial adjectives)

- *pre-approved*
- *pre-colored*
- *pre-existing*
- *predefined*
- *predetermined*

pre- + noun > nouns:

- *pre-oxidation*
- *preconsumer*
- *pre-sorting*
- *pre-disinfectant*

pre- + verb > verbs:

- *pre-dry (predried)*
- *preheat (preheated)*
- *predate (predated)*
- *prefire (prefired)*

The central meaning of *pre-* is 'before' or 'in advance'. Adjectives with *pre-* are often built with participial adjectives, nouns can be names of processes or agents. The nouns are often found in attributive position. Due to the typicality of the scientific texts of which the corpus is made up, most verbs are found in the past participle form which is part of the very frequent passive constructions. A particularity of *pre-* is furthermore that verb-constructions like *pre-cut* are used in premodifiying position in front of nouns (*pre-cut lumber*).

Adverbs of English origin were not attested.

post- 580x in ENGICOR

post- + adjective > adjectives: (including participial adjectives)

- *post-deinked*
- *post-industrial*
- *post-thermal*
- *post-treated*

post- + noun > nouns:

- *postclosure*
- *postconsumer*
- *post-injection*
- *post-monitoring*
- *post-treatment*
- *post-separation*

The majority of derivations beginning with *post-* are nouns. There are few adjectives, some of them made up of *post-*+present participial adjective. Derivatives in *–ing* are only verbal nouns used attributively (*post-recycling process, post-composting unit*). Most *pre-*+noun constructions occur in attributive use in the texts. With some words, *pre-* and *post-* are antonyms like in *pre-* and *post-mining* or *pre-* and *post-consumer*. In contrast to *pre-*, *post-* does not form any verbs. It is furthermore noticeable that *post-* occurs 79 times written as a single word with no hyphen (*post flotation, post consumer, post industrial*, etc.). The meaning load expressed by *post-* is a time-related 'after' which can also be observed in its use with years e.g. in *post-1980*.

But as with *pre-*, the corpus reveals one example of where *post-* attaches to a verb to form a premodifying construction (i.e. *post-sort conveyor*).

intra- 11x in ENGICOR

intra- + adjective > adjectives:

- *intracellular*
- *intracontinental*
- *intragenerational*
- *intravenous*

intra- + noun > nouns:

- *intranet*
- *intraplant*
-
- *intraplume*

The meaning of the prefix *intra-* is 'between' mostly in a spatial, rarely in a temporal sense (*intragenerational*). Bauer/Huddleston claim that *intra-* only attaches to adjectives (cf. 2002:1686) but the corpus reveals three examples of *intra-*+nouns which are all used attributively although at least pertaining to *intranet* the use as a full noun is well-known.

inter-

inter- + adjective > adjectives: (including participial adjectives)

- *interactive*
- *interbedded*
- *interfacial*
- *intergenerational*
- *intertidal*

inter- + noun > nouns:

- *interaction*
- *interagency*
- *intercell*
- *interface*
- *internet*

inter- + verb > verbs

- *interact*
- *interconnect*
- *interlock*

inter- + adverb> adverbs:

- *interactively*
- *interchangeably*
- *internationally*

Here again, the prefix has been used in Latin word-formation so it cannot be satisfactorily distinguished between native word-formation and loan words.

Inter- can have temporal (*intergenerational, intertidal*) or spatial meaning (*intercell, inter-agency*). Sometimes the more specific meaning of mutuality comes into play (i.e. *in interaction, interconnect* and *interlock*).

Some of the *inter-* forms contrast with *intra-* (*internet – intranet, intergenerational – intragenerational*).

mid- 213x in ENGICOR

mid- + adjective > adjectives:

- *mid-western*

mid- + noun > nouns:

- *mid-Atlantic*
- *mid-West*
- *mid-August*
- *midday*
- *mid-year*
- *mid-height*
- *midlevel*
- *midplume*
- *midstream*

Mid- can be used temporally (*mid-August, mid-year*) or spatially (*midplume, midheight*, geographic use: *mid-Atlantic, mid-West*) having the specific meaning of 'in the middle of'. Adjectives seem to be much rarer than nouns but the overall frequency of *mid*-constructions is very low. It is noticeable that *mid-* is used 73 times with time periods like *mid-1880s, mid-1940s* or *mid-50s* as well as particular years like *mid-2000.*

sub-

sub- + adjective > adjectives:

- *subaqueous*
- *sub-marginal*
- *sub-oxic*
- *sub-terrestrial*

sub- + noun > nouns:

- *subsurface*
- *subarea*
- *subcontractor*
- *subcycle*
- *submicron*
- *subsample*
- *substandard*

sub- + verb > verbs:

- *subdivide (subdivided)*
- *subcontract (subcontracted)*

Sub-, according to ENGICOR, can be attached to adjectives, nouns and verbs, displaying the central meaning of 'under' used here also for classification purposes (*subcycle, subsample, subcontract*) and indicating a more evaluative meaning of 'less than' (*submicron*). The verbs mostly occur as past participles in passive constructions.

5.2.6.1.3 Prefixes of degree and size

super-

super- + adjective > adjectives: (including participial adjectives)

- *superatmospheric*
- *supersaturated*
- *super-arid*
- *super-absorbent*

super- + noun > nouns:

- *supermarket*
- *superstore*
- *supersaturation*

Super- can be used spatially but in ENGICOR it more often bears the evaluative meaning of 'more than' or 'very'.

hyper- 60x in ENGICOR

hyper- + adjective> adjectives: (including participial adjectives)

- *hyperexcitable*
- *hyperaccumulating*

hyper- + noun > nouns:

- *hyperaccumulator*
- *hyperaccumulation*
- *hyperactivity*
- *hyper-filtration*
- *hypersensitivity*
- *hypertext*

hyper- + verb > verbs:

- *hyperaccumulate*

Most of the instances with hyper- involve members of the *hyperaccumulat** word family which pertain firmly to the new technological development in the field of soil remediation by the help of plants. In the above examples, *hyper-* bestows in most cases the meaning of 'extremely' (*hyperexcitable, hyperaccumulator*, etc.), just once the meaning of 'over' or 'beyond' as in hypertext.

ultra- 157x in ENGICOR

ultra- + adjective >adjectives:

- *ultrafine*
- *ultrapure*
- *ultramafic*
- *ultrasonic*
- *ultraviolet*

ultra- + noun > nouns:

- *ultramicrobacteria*
- *ultrafiltration*
- *ultrasonics*
- *ultrasonication*
- *ultrafiltering*

Ultra- implies the meaning of 'going beyond' (*ultraviolet, ultrasonic*) and is also used for intensifying purposes (*ultraefficient, ultrafine, ultramafic*). It combines with adjectives and nouns only. Most of the nouns with *ultra-* are names of processes.

semi- 194x in ENGICOR

semi- + adjective >adjectives: (including participial adjectives)

- *semi-volatile*
- *semi-analytical*
- *semi-annual*
- *semi-aqueous*
- *semi-automated*
- *semi-reinforcing*

semi- + noun > nouns:

- *semi-batch*
- *semi-conductor*
- *semi-crisis*
- *semi-solids*
- *semi-VOCs*

semi- + adverb > adverbs:

- *semi-annually*
- *semi-monthly*
- *semi-permanently*

Semi- sometimes has a relatively high token-frequency with one base (like *semi-volatile*) but is usually used with many different bases to create prefixed words (like *semi-batch, semi-positive*). It attaches to adjectives, nouns and adverbs and carries the meaning of 'half' or 'to some extent'. It is particularly often used with adjectives and participial adjectives.

5.2.6.1.4 Miscellaneous

anti- 74x in ENGICOR

anti- +adjective > adjectives: (including participial adjectives)

- *anti-foaming*
- *antiseptic*
- *anti-static*

anti- + noun > nouns:

- *antibody*
- *antidegradant*
- *antioxidant*
- *anti-pollution*
- *antirattlers*
- *anti-privatization*

Anti- is mostly used in the sense of 'against' or 'instead'. There are some examples of *anti-* adjectives but some more *anti-*+noun constructions which are nevertheless in most cases used attributively. Moreover, there are some attributive phrase constructions like *anti-lock, anti-skid, anti-wear* or *anti-freeze.*

self- 164x in ENGICOR

self- + adjective > adjectives:

- *self-adhesive*
- *self-cleaning*
- *self-contained*
- *self-paced*
- *self-supporting*
- *self-sustaining*

self- + noun > nouns:

- *self-absorption*
- *self-disclosure*
- *self-insurance*
- *self-weight*

Self- attaches to adjectives in particular and to nouns which are mostly used attributively. It expresses the idea of something doing something by itself (*self-cleaning, self-adhesive*) or 'being sufficient in itself' (*self-weight*). Again the corpus yields attributive phrases like *self-audit (manager)* or *self-clean (solids).*

mis- 66x in ENGICOR

mis- + adjective > adjectives: (including participial adjectives)

- *mis-colored*
- *misrepresentative*

mis- + noun > nouns:

- *misalignment*
- *miscalculations*
- *misrepresentation*
- *misuse*

mis- + verb > verbs:

- *mislead*
- *misplace*

mis- + adverb > adverbs:

- *misleadingly*

Mis- stands for 'badly' or 'wrongly' and attaches to a few adjectives, some nouns, and verbs, mostly participles in predicative position, to infer a pejorative sense. ENGICOR yields one instance of an adverb which is derived from the quite frequent participle misleading.

5.2.6.2 Suffixes

The following Table (23) gives an overview of the most used noun-, adjective-, verb- and adverb-forming suffixes together with information on what elements the suffixes combine with, examples and their most prominent meanings. This overview will not contain frequency numbers for the single affixes as it was felt to be too time-consuming to consider each single item with the particular suffix in terms of whether it belongs to native word-formation or whether it is of foreign, mostly French origin.

The use of present and past participles as participial adjectives was not included in the table but should not go unmentioned. They are a very frequent means of forming premodifying adjectives mostly derived from verbs (i.e. *abandoned site, perforated pipes, activating agent, flushing system*).

Furthermore, adverbs in *–ly* were also not explicitly included in the suffix overview on grounds of their high frequency and more inflective character.

Tab. 21: Overview on suffixes

suffix	base forms	derived form	meaning(s) acquired
NOUNS			
-age	verbs like seep, cover, leak and drain nouns like haul and short	seepage, coverage, breakage, cleavage, drainage, haulage, leakage, pumpage, package, sewage, shrinkage, slippage, spillage, tillage, shortage, storage warpage, wastage	action or result of verb action or result of noun+-ing
		mileage, tonnage, voltage	measure
-ance/-ence	verbs like absorb, allow, avoid, comply, exceed, perform	absorbance, allowance, appearance, avoidance, compliance, conveyance, disturbance, exceedance, guidance, hindrance, performance	action or state of verb+-ing
	adjectives like conform, distanct, reluctant, opalescent	conformance, distance, dominance, reluctance, obsolescence, opalescence	state of being adjective
-ing	verbs like bag, blend, sort, clean, engineer, farm, sample	accounting, aging, bagging, blending, recycling, sorting, composting, cleaning, clogging, engineering, farming, filling, grinding, mining, processing, sampling, screening	action/instance/result of verb+-ing
	verbs like build and case	building, casing	something that one verb+-s or has verb+-ed
-ment	verbs like abandon, catch, develop, embank, require	abandonment, adjustment, catchment, containment, development, embankment, impoundment, measurement, treatment, requirement	action/instance/result of verb+-ing
-ity	adjectives like alkaline, conductive,	alkalinity, availability, conductivity, density,	state or quality of being adjective

	dense, leachable, porous, toxic	electricity, feasibility, heterogeneity, humidity, leachability, permeability, porosity, toxicity	
-ness	adjectives like abrasive, bulky, effective, rough, thick, white	abrasiveness, awareness, brightness, bulkiness, coarseness, competitiveness, effectiveness, hardness, roughness, stiffness, thickness, whiteness	state or quality of being adjective
-ion	verbs like absorb, benefiate, degrade, float, neutralize, oxidize, remediate, saturate, winterize	absorption, accumulation, beneficiation, chlorination, contamination, degradation, erosion, flotation, infiltration, neutralization, oxidation, pollution, remediation, restoration, saturation, winterization	action/instance of verb+-ing
-er/or	verbs like buy, demanufacture, install, programm	buyer, consumer, demanufacturer, developer, installer, manager, manufacturer, programmer, coordinator, operator	person who verb+-s
	nouns like landscape, law	landscaper,lawyer	person concerned with noun
	verbs like analyze, buff, clarify, extrude, heat, oxidize, scrub	analyzer, baler, buffer, clarifier, composter, damper, extruder, frother, heater, liner, mixer, oxidizer, pelletizer, scrubber, aerator, reactor, compactor, evaporator, flocculator, separator	instrument/device used for verb+-ing
	verbs like bind, fertilize, fill, thicken	binder, fertilizer, filler	agents used to verb
-ist	nouns like biology, conservation, hygiene, technology	biologist, chemist, conservationist, geologist, hygienist, scientist, technologist	person concerned with noun
-ant/-ent	verbs like apply, consult, participate	applicant, consultant, participant,	person who verb+-s
	verbs like coagulate,	coagulant, contaminant,	something used for verb+-

	cool, flocculate, oxidate, pollute	coolant, disinfectant, flocculant, lixiviant, oxidant, pollutant	ing
-ics	adjectives like acrylic, organic, synthetic, plastic, toxic	acrylics, aliphatics, (geo-)synthetics, (in-) organics, plastics, toxics	classes of compounds or substances that are adjective
	adjectives like dynamic, economic, kinetic, hydraulic	(aero-/thermo-)dynamics, analytics, economics, hydraulics, kinetics	names of sciences
-ate	latin origin (plumbum acetatum)	acetate, bromate, carbonate, lactate, nitrate, phosphate	chemicals
	latin origin > analogy (i.e. precipitate > leachate)	leachate, filtrate, precipitate, condensate, injectate, raffinate	substances
ADJECTIVES			
-ive	(mostly latinate) verbs like remediate, collect, disperse, precipitate, restrict, vegetate	remediative, adsorptive, collective, degradative, dispersive, evaporative, precipitative, reactive, restrictive, vegetative	tending to verb
-able	verbs like adjust, degrade, leach, oxidize, permeate, recycle, renew, sorb	adjustable. applicable, degradable, extractable, leachable, measurable, oxidizable, permeable, recoverable, recyclable, refillable, renewable, sorbable	capable of being verb+-ed
	float, flow	floatable, flowable	able to verb
-al	nouns like accident, bacterium, cylinder, friction, mechanics, physics, technology	accidental, additional, bacterial, chemical, cylindrical, environmental, frictional, horizontal, mechanical, microbial, physical, technological	pertaining to noun/related to noun
-ous	nouns like danger, gas, hazard, mountain, odor, pore, volume	dangerous, gaseous, hazardous, mountainous, odorous, porous, voluminous	full of noun/characterized by noun
	chemical nouns like carbon, ferrus l. (iron), nitre, hydro gr. (water)	carbonaceous, ferrous, nitrous, hydrous	containing chemical noun

-y/ey	nouns like clay, glass, oil, rubber, sand, storm, water, wood	clayey, dusty, glassy, moldy, oily, rocky, rubbery, sandy, slimy, stormy, watery, weighty, woody	full of noun/covered with noun/having the quality of noun
-ic	nouns like ion, atmosphere, base, climate, electron, humus nouns like acid, aroma, ferrus l. (iron), phenol, sulfur	(an-)ionic, asphaltic, atmospheric, atomic, basic, catalytic, climatic, cometabolic, electronic, geologic, humic, volcanic acidic, acrylic, aromatic, carbonic, ferric, nitric, phenolic, sulfuric	pertaining to noun pertaining to chemicals (like names of oxygen acids and others having a high degree of oxidation)
-ish	adjectives like brack, brown, flat, gray, whitish, yellowish nouns like slug	brackish, brownish, flatish, grayish, sluggish	somewhat x somewhat like a x
-ful	nouns like insight, use nouns like harm, stress, waste	insightful, useful harmful, stressful, wasteful	full of noun causing noun
-less	nouns like bottom, color, gravel, motion, pump, trench	bottomless, cohesionless, colorless, dimensionless, gravelless, harmless, motionless, odourless, pumpless, separatorless, trenchless	without/having no noun
-like	nouns like acids, rock, soil, tree, wood	acid-like, clay-like, rock-like, soil-like, tree-like, wood-like	resembling
-type	adjectives like dry, laminated nouns like porphyry, jaw, reactor, sewer	dry-type, laminated-type porphyry-type, office-type, jaw-type, prevention-type, reactor-type, sedimentary-type, sewer-type, valley-type	like x/ resembling x
ADVERBS			
-wise	nouns like patch, step nouns like ecology, length, mileage,	patchwise, step-wise ecologywise, lengthwise, mileagewise,	in the manner of noun/like noun concerning/regarding noun

	piece, time	patchwise, piece-wise, timewise	
-wards	particles like out, down, back, up	outwards, downwards, backwards, upwards	in the direction indicated by particle
VERBS			
-ate	combines mostly with latinate noun bases	aerate, evaporate initiate, concentrate, contaminate, operate, calculate, calibrate, attenuate, differentiate	provide with noun various senses
-ify	adjectives like dense, solid, simple nouns like acid, ni-tre, gas	acidify, solidify, deni-trify, densify, gasify, specify, simplify, verify	make (into) noun/adjective
-ize	nouns like character, energy, fiber, vapour adjectives like granular, volatile, initial, industrial	granularize, operation-alize, volatilize, initial-ize, characterize, ener-gize, fiberize, industrialize, ionize, materialize, rubberize, standardize, vaporize, oxidize	make (into) noun/ make adjective/ provide with noun/perform noun
-en	adjectives like broad, damp, flat, loose, soft, sweet, tight nouns like height, length, strength, threat	broaden, dampen, deepen, flatten, lessen, loosen, shorten, soften, straighten, sweeten, thicken, tighten, weaken heighten, lengthen, strengthen, threaten	make (into) x

Looking at the sample derivational items in the table from an intuitive point of view, most of them clearly seem to belong to the domain-specific language of environmental engineering English. This is mostly due to the meaning load their base-constituents contribute like *drain(-age), absorb(-ance), leach(-ability)* or *abrasive(-ness).* What the table does not tell is which suffixes are particularly used in environmental engineering English. The question is whether their use of affixes are characteristic of a specific language too. As the table only contains examples and quantitative information of any kind and only a little qualitative information, we have to look for further information to make statements about the extent of use of the single affixes.

I will look into the productive behaviour of a few selected adjectival, nominal and verbal suffixes. The formation of nouns and adjectives in particular are of interest for a specific language as one of the characteristics of academic texts is the condensed mode of expression by using noun phrases as well as the frequent use of adjectives in order to express precise notions.

5.2.6.2.1 Productivity of Suffixes

Adjectival suffixes

In the Longman Grammar (based on the Longman Spoken and Written English Corpus) Biber et al. state that compared to other registers such as fiction, conversation and news, "derived adjectives are by far most common in academic prose" (20:531). Biber et al. use the following method to determine whether language items are productive or not: they establish their type and token frequency. An item is considered productive if it displays high type and token frequencies and if it can be established that it was used to form rare items in the recent past.

So concerning adjectives, they claim that the suffix *–al* is "overwhelmingly more common than adjectives formed with other derivational affixes", *-ent, -ive* and *–ous* to be "moderately common" and *–ate, -ful, -less, -like* and *–type* to be "relatively rare" (20:531).

The type and token frequencies for some selected suffixes in ENGICOR are shown in the table below. Note that the numbers include all instances of adjectives with a given suffix. The exact etymological determination of which items to include and which not is too time-consuming for the purposes of this study. So in terms of type and token frequency, the claims made by Biber at al. can be confirmed for ENGICOR.

Tab. 22: type and token frequency of adjectival suffixes

suffix	type frequency
-al	420
-ic	342
-able	244
-ive	159
-ous	151
-y/-ey	102
-less	45
-ful	32

suffix	token frequency
-al	34,032
-ic	15,020
-able	8,043
-ive	6,959
-ous	5,077
-ful	1,057
-y/-ey	823
-less	373

Concerning their sheer type and token frequency, the items on the top ranks of the table are more important than the lower ones. The ranking established by type and token frequency look also quite similar, particularly for the top ranks. As establishing the productivity is just a means to determine their importance in the teaching processes, it can be stated that concerning text coverage *–al* adjectives should be focused on more than adjectives in –less as in ENGICOR every 59th word ends in *–al* compared to every 5,362nd in *–less*.

As the validity of information based on whether language items occur in the OED or not and if so how many of them are 20th century formations has already been established in the 5.2.5.2, I will also use them for derivational forms in order to amend the purely quantitative examination above.

Tab. 23: occurrence of adjectives in OED

suffix	**number of types not listed in the OED**
-able	20
-ic	16
-less	9
-al	6
-ive	2
-y/-ey	0
-ous	0
-ful	0

suffix	**number of types listed after 1900**
-ic	52
-al	23
-able	14
-ive	5
-less	4
-ous	4
-y/-ey	2
-ful	1

This table shows quite a different picture in detail but the overall tendency from the type and token frequency, i.e. that *–al*, *-ic*, and *-able* are more productive than *–ive*, *–y/-ey*, *-ous*, *-less* and *–ful* is confirmed.

-able has the most items not listed in the OED, followed by *–ic*, *-less* and *–al*. A closer look at the not listed items in –able discloses that rather than being obsolete and rare formations, most of them are innovations in the field of environmental engineering English.

Tab. 24: Adjectives not listed in OED

-able	
autoclavable	trommelable
baleable	inhaleable
contactable	maneuverable
compactable	tailorable
configurable	trommelable

demountable	atomizable
depletable	exposable
emulsifiable	compostable
flockable	flowable
fusable	sorbable

-ic, -al and *–able* have a relatively high number of neologisms listed in the OED after 1900, so they have proved to be productive throughout the 20th century and are very likely to be ready to form new items in the future as well. The high number of *–ic* derivatives is also due to the ability of the suffix to form chemicals-related adjectives.

Considering the lower type and token frequency of *–able* compared to *–al* and *–ic*, this suffix seems to catch up and be more frequently used in the formation of new words in environmental engineering English at the moment.

Having established *-ic, -able* and *–al* as our favourites in terms of expected specific vocabulary growth, a short look at the productivity in the narrow sense of the concerned items is enough to confirm the view established in the chapter on combining forms that the nature of data in such a small corpus is not sufficient to serve as an indicator for productivity.

Low frequency items like *–y/-ey, -ful, -less* and *–ous* gain too much weight on grounds of their high number of hapaxes compared to their low token frequency. So the results are almost in reverse to the other rankings.

Tab. 25: productivity in the narrow sense of adjectical suffixes

suffix	**productivity in the narrow sense**
-y/-ey	0.0473
-less	0.0348
-ful	0.0113
-ous	0.0096
-able	0.0084
-ic	0.0061
-ive	0.0031
-al	0.0025

Nominal suffixes

The same procedure was tried with a selection of nominal suffixes: *-ness, -ity, -ion, -ant/-ent* and *–ist,* see Table 28:

Tab. 26: type and token frequency of noun suffixes

suffix	number of types
-ion	610
-ity	271
-ness	102
-ant/-ent	74
-ist	36

suffix	number of tokens
-ion	63,672
-ity	19,166
-ant/-ent	11,730
-ness	1,860
-ist	244

In terms of text coverage, *-ion* ranks highest in type and token frequency, which makes *–ity* with about a third of the numbers a poor second. So every 31st word in the corpus ends in *–ion*. The switching in the order between *–ness* with a higher type frequency and *–ant/ent* with a higher token frequency is interesting.

If we follow the opinion of many linguists that a high amount of low-frequency items is a sure sign for a high productivity and a high amount of high-frequency items a sure sign for low frequency, the ability of *–ness* to produce neologisms should be much higher than of *–ant/-ent*.

This is where we need to look at the listedness of the concerned items in the OED:

Tab. 27: occurrences of nouns in OED

suffix	number of types not listed in the OED
-ity	17
-ion	8
-ant/-ent	6
-ness	1
-ist	0

suffix	number of types listed after 1900
-ity	34
-ion	34
-ant/-ent	11
-ist	3
-ness	1

These numbers suggest a different picture again: whereas 6 nouns in *–ant/-ent* are not listed in the OED, it is just one which ends in *–ness*. Moreover, there are 11 nouns in *–ant/-ent* listed in the OED after 1900 and again just 1 ending in *–ness*.

A look on an even more concrete level confirms the assumption started by the sheer numbers that *–ant/-ent* produces more new words in environmental engineering English than *–ness*.

Tab. 28: occurrence of nouns in OED

words in –ness not listed in the OED	words in –ness after 1900	words in –ant/ent not listed in the OED	words in –ant/ent after 1900
abrasiveness	uneveness	anti-degradant	eluent
		prop(p)ant	suppressant
		titrant	coolant
		interferant/interferent	extractant
		flocculant	sealant
		lixiviant	adsorbent
			retardant
			supernatant
			reactant
			reductant
			surfactant

With the help of information from the OED, *-ity* can be established as the most productive suffix in environmental engineering English followed by *–ion* whose productivity must be considered much lower given its high token frequency. The data furthermore confirm a further claim made by Biber et al. in the corpus-based Longman Grammar of Spoken and Written English: the predominance of *–ity* in academic prose over its rival suffix *–ness* (cf. 20:325). Biber et al. justify this phenomenon – which is the reverse in General English – by two particularities of academic prose:

1. the ability of *–ity* to combine with foreign bases which are more common in academic prose
2. the tendency of *–ity* to denote abstract entities and abstract relations in contrast to –ness which rather denotes qualities of persons and things (cf. Biber et al. 20:325)

Verbal suffixes

In the following table those verb lexemes ending in *–ize, -ate* and *–ify* which are either not attested in the OED or are attested with an entry date after 1900 are collected. The verbal suffix *–en* only attaches to a very small number of bases and is no longer productive and is therefore not included in this investigation.

Tab. 29: occurrence of verbs in OED

types not listed in the OED			types listed after 1900		
-ize	-ate	-ify	-ize	-ate	-ify
commoditize	dehalogenate	viscosify	aerosolize	fluorinate	ammonify
compatibilize	evapotranspirate		anodize	hydroxylate	detoxify

fiberize	ozonate		channelize	passivate	
granularize	chelate		compartmentalize	protonate	
hydrolize			computerize	phosphate	
incentivize			containerize	solvate	
levelize			customize		
palladize			pressurize		
rubbleize			digitize		
			finalize		
			hospitalize		
			initialize		
			micronize		
			motorize		
			operationalize		
			palletize		
			pelletize		
			plasticize		
			prioritize		
			privatize		
			publicize		
			rubberize		
			solubilize		
			traumatize		
			winterize		

It can be stated that when applying the historical-dictionary-based method to establish the likelihood of forming new words, *-ize* seems to be the most likely candidate. *–ate* can be considered slightly productive and *–ify* only produces the occasional new word.

By means of the suffix *–ize* verbs conveying different meanings are formed: *aerosolize > to become aerosol, rubbleize* and *fiberize > to make into rubble, fiber*, *containerize, computerize and hospitalize > to put into a container, onto a computer, into a hospital*, and *hydrolize* and *palladize > to provide with hydrogen and palladium. –ize* combines with bases of foreign and native origin; with adjectives and nouns.

-ate predominantely forms verbs that describe chemical processes where substances are either added (*to phosphate, to ozonate, to chelate*) or substances get formed (i.e. *to solvate*). Non-chemical words have a similar meaning: *to render passive = to passivate, to render evapotranspirative = to evapotranspirate.*

The few instances of new words formed by the help of *–ify* have the meaning of undergoing a process: *detoxify = undergo detoxification, viscosify = become viscous, ammonify = undergo ammonification.*

5.2.7 + Omission

5.2.7.1 Initialisms

It is especially in science and technology that abbreviating longer terms is a very popular way of shortening multiword names, which they replace as a short and concise version of the sometimes lengthy originals. This is done by usually taking the first letter of the words that make it up. There are two ways of pronouncing the resulting item: as with abbreviations the letters are pronounced separately, like *AMD (acid mine drainage)* or *BOD (biological oxygen demand)*, or as with acronyms, the letters are pronounced in a sequence "having their characteristic phonological value" (Bauer/Huddleston 2002:1633), like *NAPL (non-aqueous phase liquid)*. One example from the corpus is a mixture of the two: *DNAPL*, where the initial *D* is pronounced as single letter followed by *NAPL* pronounced in a sequence [næpl], which is about the only way to make this item pronouncable.

20 initialisms feature among the 500 examined keywords from the comparison of ENGICOR with the BNC. These 20 items I take to illustrate this phenomenon, although the corpus as a whole contains more. They are all highly specific to the domain of environmental engineering and designate mostly noun compounds and noun phrases naming special terms. Most of them are written with capital letters, the only exception is *pH.*

None of the chosen examples fits the pattern described by Plag where the items are "no longer related to the words they originally abbreviated" (2003:127), like for instance *radar*, although users might well have difficulties recalling every part of the originals correctly.

In terms of behaviour in the morphological field, some of the initialisms can inflect for the plural (*DNAPLs, VOCs – volatile organic compound(s), PRBs – permeable reactive barrier(s)*), and half of them get involved in derivational and compounding word-formation processes: there are a few prefixed initialisms like *post-UV* and *pre-UV (samples)* and *semi-VOCs*. Compounding seems to be a bit more frequent and varying: there are mostly complex adjectives produced by combining either (1) an initialism with an adjective (i.e. *UV-sensitive, UV-specific, VOC-free*), (2) an initialism with a past participial adjective (i.e. *AMD-related, PCE-/DNAPL-/TCE-contaminated, UV-treated, VOC-laden*) or (3) an initialism with a present participial adjective (i.e. *PCE-containing, UV-absorbing, VOC-emitting*).

There are furthermore a few initialism + noun compounds like *LFG-collection (landfill gas), UV-light* or *VOC-treatment* which are in most cases used in attributive position and only one is used in a phrasal compound (*LFG-to-energy recovery systems*) where it becomes particularly obvious why we use initialisms: the full word *landfill-gas-to-energy recovery systems* would be much longer.

Concerning their internal make-up, nouns are the most frequent building stones for the examined initialisms, followed by adjectives; prepositions, for instance, play no role amongst the 20 most frequent initialisms in the corpus.

Most of the 20 examples fit the pattern adjective modifying a noun+noun compound like *AMD (acid mine drainage), COD (chemical oxygen demand)* and *TKN (Total Kjeldahl Nitrogen)*, while others have an adjective modifying an adjective+noun compound like *VOCs (volatile organic compounds), MSW (municipal solid waste)* and *GAC (granulated activated carbon)*. In another example an adjective+noun combination modifies a complex noun, i.e. *HDPE (high-density polyethylene)*, others are abbreviations for chemical compounds, i.e. *PCE (perchloroethylene)* and *TCE (trychloroethylene)*.

In *TCLP* the density of information seems particularly high: it is the abbreviation for *Toxicity Characteristic Leaching Procedure*. The only abbreviation that is word-class-wise an adjective amongst all the nouns is *UV (ultraviolet)*.

5.2.7.2 Clippings

When parts of one word are cut off at the beginning, at the end or at the beginning and the end, the results are called clippings. It is also the clipped part which serves as the prerequisite for classifying clippings; those whose endings are cut off are called back-clippings. They are the most frequent ones (cf. Bauer 1983:233). When the beginning of a word is clipped, we speak of a front-clipping and getting rid of the beginning and the end produces ambiclippings.

The corpus yielded just two findings by chance: *lab* from *laboratory* and *moly* from *molybdenum*, both backclippings. This comes to no surprise as the use of clippings in language is more common in a situation where the participants in the communication process know each other well. According to Fischer (1998:40) many clippings are stylistically marked or like Plag remarks "used to express familiarity" (2003:117). But as the corpus consists of mainly written texts, there is a certain distance between the writer and the readers which hinders the use of more clippings whereas I assume we can expect to see more of them. Moreover, *lab* has become quite commonly used and is used as a simplex as well as in compounds (*lab analysis, lab test, lab studies*) in order to achieve language economy. The motivation why the author used *moly* in the compound *moly-plant* in a formal technical resource document remains opaque. *Molybdenum plant* features four times in the text, *molybdenum* 44 times altogether, and there is just one instance of *moly*. With the concrete sample sentence in mind (*... tailings are generated during the initial copper-molybdenum flotation circuits, prior to the separation of the molybdenum from copper in the moly-plant.*), one is only tempted to think that the author wanted to avoid saying *molybdenum* for the third time in the same sentence.

As it is not possible to search ENGICOR systematically for clippings, further examples might have excaped my attention.

5.2.7.3 Backformation

New words are created by processes called backformation when affixes of morphologically complex words are deleted. Usually the speaker actively involved in this process might at least implicitly be aware of an "identical relationship from a synchronic point of view" (Quirk 1985:1578), i.e. with the corpus finding of *oxic* from *anoxic* which were thought to relate to each other in the same way as *aerobic* to *anaerobic*. Whereas in the latter example *anaerobic* derived from *aerobic* by the addition of a prefix, *oxic* was formed when the suffix was taken off from *anoxic*. That is what caused Bauer to call backformation a process where the "rule of word-formation is reversed" (1983:231).

According to Bauer (cf. 1983:230) and Quirk (cf. 1985:1578/79) the majority of backformations are verbs and Bauer/Huddleston add that "(b)ackformation by the deletion of –ing or –er is indeed one of the main avenues for the creation of verb compounds with the structure noun base+verb base: baby-sit,..." (2002:1637) This phenomenon has already been mentioned above in the chapter on compounding concerning assumed compound verbs with a verb as a second element. Here it was found that rather than being a result of a compounding process, items like *size-reduce* from *size reduction*, *starve-feed* from *starve-feeding* and *freeze-dry* from *freeze-drying* are more likely to be backformations. From these examples it becomes clear that the motivation to use such backformations must be to avoid more complex phrases which would normally be involved to express the same meaning (*to reduce in size, to feed an extruder at a rate below the full capacity, to freeze by drying at the same time*).

Similarly, looking at their first mention in dictionaries, the noun *adsorption* existed first before the verb *adsorb* was backformed; *sorbent* developed by deleting the prefix from *adsorbent, flocculant* by deleting the de- from *deflocculant*.

5.2.8 +Combination/+Omission

5.2.8.1 Blends

Intensive contact with the corpus resulted in findings of some examples of blending processes purely by chance. A blend is a lexical item that results from the combination of at least two words of which at least one is shortened. In Fischer (1998:34) we can find the description "clipped compounds" for a blended item which points to the facts that blends resemble clippings in the sense that parts of words get cut off and compounds in that parts of words get combined.

Statements about the transparency of the morphological components of blends differ; whereas Quirk says that "(e)nough of the each (word) is normally retained so that the complex whole remains fairly readily analysable." (1985:1583). Bauer claims that "the etymological root of the word is only clear when specifically explained" (1983:234). The reality confirms both statements: some blends are easily understandable like *glasphalt*, others certainly need further explaining like *mafic*. In Lehrer this phenomenon is reflected when she tests the following hypothesis successfully: "the more material from the target word is present, the easier the blend is to identify" (1996:366).

Eight blends were detected by chance in the corpus which does not allow a comprehensive categorisation but a small exemplification of their characteristics. In blends it is not only forms that combine but also the meanings in order to fulfil the task of using language economically. *Soilcrete* is grouted soil, *glasphalt* is asphalt with a high proportion of glass, *fertigation* is the spreading of fertilizer by irrigation systems. The resulting language items moreover require a special and additional meaning as they are used in new contexts.

From a formal point of view Bauer/Huddleston (2002:1636) distinguish four types of blends which I will try and illustrate with the corpus findings. The first type comprises blends which are formed by the first part of the first base and the whole of the second: in the corpus we find *evapotranspiration (evapo(ration)+ transpiration)* which designates the combined process of evaporation from the soil and transpiration of the plants. The second type has the whole first part of the first base and the final part of the second word: i.e. *soilcrete (soil + (con)crete))* fits this type. In the third type, we find the first part of the first base and the final part of the second base: *pervaporation (per(meation) + (e)vaporation)* can be named here. And in the fourth type there is an overlap between the two bases in the central part of the blend: *fertigation (ferti(lizer)+(irr)igation)* could fit this category as it cannot be said for sure where the central *"i"* belongs to, and *glasphalt* (the product obtained when recycled glass is added to asphalt) is at least phonologically seen a member of this category.

Some of the examples that I put into the blends category do not only involve blending but also derivational processes: *mafic* consists of *ma(gnesium)* + *f(errum)* and the adjective suffix *–ic*. *Surfactant* means *surf(ace)+act(tive)* and the noun suffix *–ant* designating agents.

Blending is an intentional process (cf. Marchand 1969:451) which the following OED sample sentence for surfactant reads proves: "A new word, Surfactants, has been coined by Antara Products; General Aniline & Film Corporation, and has been presented to the chemical industry to cover all materials that have surface activity, including wetting agents, dispersants, ..." (OED-entry for surfactant). According to Detering (2000:89) blending is common in commercials, names of companies, institutions, projects and names of technical equipment. With only eight examples that came to light, it is not very frequently used in environmental engineering English (although I cannot exclude the fact that I may have missed more examples).

Diachronically seen and supported by corpus evidence, some words that were categorised above as products of combining processes with neoclassical compounds could also result from blending. In the corpus we find *biodegradation* as well as *biological degradation, biobarriers* as well as *biological barriers, bioactive* as well as *biologically active*. Here the reduction of *biological* to *bio-* and the neoclassical compound *bio-* get mixed up not only in the minds of the environmental engineers who created and are using this term but also in the minds of linguists because the process which has actually been used, be that combination or blending, cannot unambiguously be determined.

With *eco-* the case is set more clearly: it is a shortening of *ecological*, so words like *ecosystems, ecocapitalism, eco-efficiency, eco-fees, eco-minded, eco-tax* and *eco-tourism* can be added to our original list of blends.

Evapotranspiration, pervaporation and *fertigation* are all examples for dvandva-blends where the two constituents are semantic coordinates (cf. Fischer 1998:35). If the *eco*-examples are to be considered examples of blending processes, they would account for endocentric blends with the *eco-* modifying the second element.

5.2.9 -Combination/-Omission

5.2.9.1 Conversion

In conversion processes, a lexical item which is a member of one word class is used in another word class without any change of form. In this way new words are created " on the basis of already existing ones" (Plag 2003:107). This is an "extremely productive way of producing new words in English" (Bauer 1983:226) and happens to words of any kind of word class although conversions from the main classes; noun to verb, adjective to noun, adjective to verb, and verb to noun, are the most frequent. Furthermore, the wide term lexical item is used for a purpose: not only simple words can undergo conversion processes but results from all kinds of word-formation processes can also change word class.

Before I come to look in more detail at the conversion of adjectives to nouns shown in the *–ables* case study, I want to illustrate the characteristics of conversion by the help of a few examples from the corpus. Unlike in the case of *–ables*, there is no way to systematically search for products of conversion processes and the examined examples now are found by chance.

As the means to produce verbs with the help of derivation is mainly limited to just three suffixes, conversion is in particular frequently used to produce verbs (cf. Hansen 1990:127). Moreover, the corpus yields two examples of where this has happened with a compound noun as source which produced a compound verb: *thin section > to thin-section*, and *landfill > to landfill.*

In order to find out about the direction of conversion a diachronic view has to be adapted looking at the entry dates in historical dictionaries. The noun *landfill* was – according to the OED - first mentioned in 1903 whereas the first mention of the verb was in 1988. With results of more recent developments in language like *thin section/to thin-section*, the corpus needs to serve as a source to find indications for the direction of conversion as there is no OED-entry for it. According to Plag (2003:107), the derived words are "semantically more complex" and thus "tend to have a narrower range of meaning, to the effect that they cannot be used in as many contexts as their base words" (2003:109). Consequently the base words should be more frequent in the corpus. This is confirmed when we look at *thin section* as a noun with 6 detected instances and just one instance of *thin-section* as a verb (*The sediments were impregnated with epoxy resin and thin-sectioned for electron microprobe...*). More drastic evidence is brought to light by looking at the concordance of *landfill*, separating the nominal from the verbal use: nouns occur 2385 times, verbs just a mere 67 times. The additional meaning the verb acquired is 'to take materials to a landfill', and 'to prepare rock in the form of thin sections', which in both cases exceed the semantic complexity of the original nouns.

As mentioned above, conversion happens in all word classes. The examples of *unders* and *overs* where nouns derived from preposition are evidence of that. These two items designate materials in screening processes which either pass *(unders)* or do not pass a screen due to its size *(overs)*. As we will see in the –*ables* case study below, *unders* and *overs* as well in the form of nouns are used to cover a class of materials which have a certain feature in common, i.e. 'being able to pass a screen or not'.

5.2.9.1.1 Conversion from Adjective to Noun: A Case Study

As was introduced in the previous paragraph, conversion can be considered as an instance of derivation without any change in form. In the case of a conversion from adjective to noun, the adjective form is retained whereas the resultant word form takes on the characteristics of the new word class, i.e. of nouns.

The starting point of this case study was an investigation into all nominal suffixes featuring in the corpus where amongst others the following lexical items – all displaying the ending –*ables*- happen to appear:

Tab. 30: occurrence of –*ables*-words in ENGICOR

lexical item	**corpus frequency**
recyclables	129
floatables	106
non(-)durables	12
deliverables	11
non-renewables	7
biodegradables	4
consumables	2
nonrecyclables	2
expendables	1
extractables	1
imponderables	1
moveables	1
non-refillables	1
nonusables	1
receivables	1
refillables	1
renewables	1
unrecyclables	1

Surely a marginal phenomenon considering its frequency, and yet its occurrence triggered a lot of questions about its origin, and about its extent of use in terms of belonging to a specific domain. In addition, questions arise as to whether adjectives with suffixes other than *–able* have the same ability to generate nouns by conversion, and what their semantic particularities are. To start with, I will have a look at the morphological background of these items.

Morphological Background – the Reference-Book Approach

Origin

The OED was consulted to find out about the etymological roots of present-day adjectives ending in *–able*. A look at the entry for *–able* revealed that adjectives with this suffix were first adopted from French whereas the French suffix *–able* goes back to Latin adjective suffix –ābilem which was usually added to verbs in –āre, -er in French. After an initial stage of adopting whole adjectives from French (i.e. *recevable, mouvable*), the suffix was later recognized as such and used as a fully fledged word-formation element to form native words (i.e. *bearable, speakable*). The fact that *-able* has been associated with the adjective able, to which it is morphologically not related, is supposed to have acted as a supporting factor for the spreading of these adjectives (cf. OED-entry for –able, Marchand 1969:230).

Form and Semantics

A consultation of standard textbooks of English word-formation (Quirk 1985:1555, Marchand 1969:229-232, Hansen 1990:111-112, Plag 2003:94-95) confirms the findings on the origin and adds information on the formation of –*able*-adjectives and on its semantics. The suffix *–able* can be added to transitive and intransitive verbs as well as to nouns in order to form adjectives with active meaning (capable of X+-ing, i.e. *floatable*) or even more frequent with passive meaning (capable of being X+-ed, i.e. *recyclable*). Adjectives of French origin remain opaque to the present-day speaker of English but can be analysed in the same way going back to the French or Latin roots (*durable* from the French verb durer, from the Latin verb durare).

Deverbal formations are considered more frequent than denominal ones (cf. Marchand 1969:230).

The consultation of reference works brought first insights into the development of word forms ending in *–able* in their function as adjectives. Their syntactic role is to modify nouns when in attributive position and characterize noun phrases when in predicative position. They denote qualities of classes of objects. In a next step, the corpus was consulted in order to find out which nouns are modified by the corresponding *–able*-adjectives of the lexical items that were attested as *–ables*-nouns in Table 32. In this way, it was hoped to establish frequent adjective-noun collocations which might reveal more information on the process of conversion from adjective to noun.

Tab. 31: collocates of *–able* adjectives

adjective	fre-quency	premodified noun(s)	frequency
recyclable	101	material(s)	44
floatable	38	material(s)/pollutants/solids	17/3/4
non(-)durable	47	goods/products	37/5
deliverable	/		
durable	110	goods/material(s)/products	37/3/6
non-renewable	21	resource(s)	21
biodegradable	135	(hydraulic) fluids/(organic) materials/ (organic) compounds/organic matter/ organics/product(s)/slurry/ies/ (organic) waste(s)	13/5/3/4/8/9/10/ 7
consumable	/		
nonrecyclable	8		
expendable	3		
extractable	8		
imponderable	7		
moveable	4		
non-refillable	17	((beer) and (soft drink))/(beverage) containers	17
non(-)usable	6	material	4
receivable	3	accounts receivable	3
refillable	33	(beer)/(glass)/(soft drink) bottles/(beer) containers	15/6
renewable	49	energy (sources)/resources	11/18
unrecyclable	1		

(minimum occurrences of modified noun(s)=3)

It is noticeable that the adjectives which occur in the corpus show a clear preference to collocate with one particular noun or with a group of nouns that display the same semantic preference (i.e., semantic preference of nouns collocating with *biodegradable*: = organic, mostly liquid substance that may be degraded).

Furthermore, it can be stated that the collocating nouns are some kind of collective term, very often a material, product or good. These nouns are mostly generic nouns with little semantic weight whereas the actual meaning load rests on the adjective.

Some items are not numerous enough to display obvious patterns (*expandable, extractable*). Others like *consumable* might be institutionalised to the degree that the original two-word combination is not used any more, and in the case of *moveable* which can be used with any kind of object that is not fixed, the adjective instances are only remotely related to the semantics of *moveables* which is any kind of property that can be moved. Still, these are presumptions that can only be verified with the aid of a bigger corpus as such a small amount of instances cannot serve as proof.

However, the examples which are established as both - *-able*-adjective+noun collocates and as *–ables*-noun - can be regarded as pervasive.

The found corpus data leads to a first hypothesis:

Nouns ending in *–ables* started their lives as well-established adjective+noun collocates whose noun assumed an ellipted form over the time. The adjective converted into a noun carries the same meaning as the former compound.

The reference book approach revisited

Turning again to reference works like dictionaries and grammar manuals, I will look for information on conversion processes of nouns from adjectives and whether this process is regarded as a marginal or frequent development.

First, it must be said that information of the sought kind is very scarce.

Hansen (1990:134) mentions the possibility of nominal zero-derivations from adjectives and provides the examples *empties* and *heavies*. He stresses that the resultant lexical items are often plural forms denoting persons (*black, imbecile, intellectual*) or objects (*chemical*).

Marchand (1969:359-364) describes several conversion processes but does not mention instances of nouns derived from adjectives.

It is Quirk then who gives the first more substantial hint. Here (1985:1558-1567) there is a mention of full and partial conversion processes whereas the partial conversion process is demonstrated by the help of *the wealthy* which is regarded as an "adjective (…) partially converted to noun status, in that it is syntactically in a position (head of a noun phrase) characteristic of nouns rather than adjectives" (ibid.). Still it is not recognised as an example of a full conversion in that it is not able to "behave inflectionally like a noun" (ibid.), as it cannot vary in terms of number and case.

Further on (1985:1560/61), Quirk makes another note of the adjective-noun conversion when he illustrates the process by examples such as *natural* for *naturally skilled player* and *final* for *final race*. He comments that these examples indicate "no very productive pattern" (1985:1560). Additional examples are *daily, weekly, annual, perennial, comic, regulars*, etc.

"From these examples", he reasons," it is seen that adjective>noun conversion can usually be explained in terms of a well-established adjective+noun phrase from which the noun has been ellipted" (Quirk 1985:1561).

Bauer describes "established examples of adjective>noun conversion " as "relatively rare and frequently restricted to their syntactic occurrence" (1983:230). There is no explicit mention of full or partial conversion but *the poor* is given as an example which cannot be made plural or take a determiner and *a daily, a regular* and *a roast* serve as less restricted examples. The following quotation which follows immediately after the previous examples deserves special mention: "This type seems to have become much more productive recently, and recent examples include a creative, a crazy, a double, a dyslexic, a gay, a given, an inflatable, a nasty." (Bauer 1983:230)

A few years later, in their Chapter on Lexical Word-Formation in the Cambridge Grammar, Bauer and Huddleston provide a fuller picture of the process. They describe it as development where "(i)n most cases, the resulting noun is equivalent to a nominal consisting of the adjective and understood noun,..." (Bauer/Huddleston 2002:1642). Far from attempting an exhaustive classification they mention different groups of resultant lexical items like some belonging to informal style (*dear, heavy, royal*), some denoting types of persons, colour terms, time-adjectives (*daily, weekly*) denoting periodicals, some occurring normally or exclusively in the plural (*basics, greens, marrieds, smalls, woollies*). Furthermore, they state that these items will often have two or more senses (*daily* as periodical and *daily* as a cleaner who comes daily).

Bauer/Huddleston are also the source of information for claims that language phenomena like "the poor" are clearly different from "intellectuals" (2002:418). This claim is made this time based on the fact that *intellectual* can inflect for number (*the intellectuals* with plural-marker –s) and that it can take an adjective as modifier (*remarkable intellectual*) whereas *the poor* can only be modified by an adverb (*the remarkably poor*) (ibid.).

Apart from information on its distinctness from fused modifier-heads like *the poor* and information on its semantics mentioned above, Bauer/Huddleston also refer to the conversion of adjectives to nouns as a process that comprises "(a) considerable number of nouns formed by conversion..." (2002:1642) which is the only remark on the frequency of the process.

Summing up from the retrieved scarce information from reference works, it can be assumed that the process of conversion from adjective to nouns is a marginal process that happens to some well-established adjective+noun collocations. Some semantic fields where this process happens were established (like colour terms and denotations of periodicals). Still, there is a very confused picture as to what extent this process is alive and working in language use.

It can be said that the investigated process did not always receive special mention in grammars and if it did, the descriptions were not very comprehensive.

By looking at the recurring examples given in the reference works one can assume that this is a very restricted phenomenon in a few categories.

The Exception from the Rule

Two of the above-mentioned examples already provide the ground for the next approach which differs from the aforementioned in the consulted reference works. There was reference to *a daily* which stands for a cleaner who comes daily and *a natural* for a naturally skilled player. If their origins are as described, neither of them stems from an adjective-noun conversion process but from a process where a longer phrase underwent an elliption process to a shorter form which in this case happens to be a noun with an adjective form. Raymond Hickey (2003) looks at such processes and calls them univerbation. In contrast to the more specific process of conversion, he defines univerbation as "a more general shift in the language by which phrases are reduced to single words", which "can be a verb, noun or adjective"(2003:94) His examples comprise verbs that derived from verbal phrases (*tracking* from *keeping track*), verbs from nouns (*interface* as a verb), nouns from verbal phrases (*a piss-take* from *taking the piss out of someone*) and adjectives from verbal or prepositional phrases (*a gated community* from *a community which lives behind gates*). We find our examples of adjectives derived from adjectival phrases again in Hickey's article in the discussion section (2003:102) when the motivation for such processes is looked at. The relatively simple process of the deletion of one element in the case of adjectival phrases getting shortened to nouns is opposed to more far-reaching processes as in the given examples above where "an active restructuring of the syntax" (2003:102) and "some subtle semantic shift" (2003:93) happens to take place. He claims that univerbation such as that in the case of nouns derived from adjectival phrases can happen because of the "clarity of context" (2003:102). Looking at the process of the deletion of nouns from an adjective-noun compound from this angle opens a whole new perspective. It places the –

ables-words and their like in a more comprehensive context of "an analytic language, where lexical compaction, i.e. univerbation, is in evidence..." (2003:94), and conversion is only one part of an ongoing development.

Widening the Perspective – The Corpus Approach

Next, the question whether adjective-noun conversion is a phenomenon typical for adjectives ending in *–able* or whether it can happen to other adjectives as well is in the focus of the investigation. First, typical adjective-endings like *–ive, -ant/ent, -al, -ic,* and *–ous* were chosen for reasons of good searchability by the help of corpus concordance.

Adjectives in *–ous* could be eliminated relatively early from the investigation as the only example that displayed an at least partial conversion appeared in the sentence *Don't' overlook the obvious*. So as a first conclusion, it can be stated that adjectives ending in *–ous* do not produce adjective-noun conversions.

For an account of the other endings the corpus was searched for nouns ending in *–ives, -ants/-ents, -als* and *–ics*: Amongst many others, the following couples could be detected:

- *abrasive* *abrasives*
- *additive* *additives*
- *corrosive* *corrosives*
- *intrusive* *intrusives*
- *positive* *positives*

- *adsorbent* *adsorbents*
- *effluent* *effluents*
- *flocculant* *flocculants*
- *constituent* *constituents*
- *gradient* *gradients*

- *chemical* *chemicals*
- *lateral* *laterals*
- *local* *locals*

- *physical* *physicals*
- *residual* *residuals*

- *acrylic* *acrylics*
- *analytic* *analytics*
- *economic* *economics*
- *inorganic* *inorganics*
- *hydraulic* *hydraulics*

At first sight, there seems to occur the same phenomenon as with the adjective-noun couple *–able/-ables*. Consequently, facts either for or against parallel processes have to be established. Now the question is whether the resultant nouns are a consequence of conversion or whether other word-formation processes have come into play. In order to find this out, the Oxford English Dictionary (OED) was consulted and information sought on the etymologies of the words. Attention was paid to characteristics they have in common and where differences occurred in the naming processes.

Widening the Perspective – The Reference Book Approach

Adjectives in *–ive* – Nouns in *–ive(s)*

The OED-entry for the adjective-ending *–ive* reveals the following: the "suffix is largely used in modern Romanic languages, and in English to adapt Latin words in –īvus , or form words on Latin analogies".

The examples from the corpus like *abrasives, additives, corrosives* can be detected in the OED and the OED etymology establishes that the adjectives either go back to Latin adjectives as in the case of additive, where the dictionary refers back to the Latin additīv-us , or constructs a kind of mock-Latin adjective where it is not established as in the case of explosive from *explosiv-us from the verb explōdĕre and refers at the same time to the French equivalent (explosif/-ive).

An interesting comment is found in the OED-entry for the adjective-ending *–ive* again: "already in Latin, many of these adjectives were used substantively, this precedent is freely followed in the modern Romanic languages and in English" (i.e. *adjective, captive, derivative*, etc.).

So, if still in use, the process of producing nouns with adjective-endings seems at least for this adjective ending an ancient process whereas it remains unclear if it happens by conversion or whether the process is more direct.

Hints for a more multi-facetted and ongoing process comes from the OED as well. Some *–ives* entries display the additional remark "The adj. used absol.", referring to the use of the adjective in noun position. This can be found in the entry for *adhesive* as an *adhesive substance*, of which the first occurrences were documented in 1912. What is even more revealing in the entry *objective* is there is the additional remark "elliptical use of the adjective" and we learn from the OED that the noun *objective(s)* is short for *objective point* which originates from military language of which I assume nowadays only few speakers are aware. Thus, a conversion very much like in the established *–ables* examples above must have occurred.

Adjectives in *–ant/-ent* – Nouns in *–ant(s)/-ent(s)*

Well-established adjectives ending in *–ant/-ent* can – according to the OED – be traced back to Latin (i.e. *coagulent* to coāgulānt-em, the present participle of coāgulāre) and often also referred to their French equivalent. Further examples are *entrants, disinfectants, attendants* and *applicants*.

In examples like *contaminant* and *retardant*, it is referred to a verb or noun and the adjective-ending *–ant* (i.e. *contaminate+ant, retardance+ant*). Thus, the ending must have become productive in its own right in English.

The retrieved examples again reflect that the different resultant nouns are at different stages of an on-going development. We can find adjectives like *buoyant* and *flocculent* where the OED does not provide an entry for a noun whereas the corpus shows them (*buoyants* – one occurrence, *flocculents/flocculants* - 22 occurrences). Others have neither an entry for the adjective nor for the noun (*lixiviant(s), leachant(s), solidificant(s))*. And others again are clearly the result of analogy processes like *coolant(s)* after *lubricants* (cf. OED-entry for *coolant*) and *reductant(s)* after *oxidant(s)* (cf. OED-entry for *reductant*). Corpus-findings like *anti-ozonants* and *antidegradants* are very likely to be built by analogy on the grounds of *anti-oxidants*.

One finding which merits particular emphasis is the fact that there is no absolute need for the existence of an adjective ending in *–ant/-ent*. Moreover, the ending *–ant* (more than *–ent*) seems to have developed a meaning and consequently a productive life of its own, able to produce domain-specific nouns where the *–ant* denotes some kind of agent, i.e. *sealant, extractant, suppressant, lixiviant, leachant, anti-ozonant, anti-degradant*, etc.). These words can be regarded as significantly different from the old forms like *applicant* and *incident*.

The case of *surfactant*, which is a blend of the multi-word item *surface-active agent*, can be regarded as further proof for this hypothesis, where the blend acquires the ending and displays the same meaning "agent" as established in the long form.

Adjectives in *–al* – Nouns in *–al(s)*

Looking at the OED-entry for *–al* as adjective-ending reveals a by now familiar scenario: we are referred back to a Latin origin, the Latin adjective suffix –āl-em and its long journey through French is mentioned as well. Here as well, we find mention of the possibility of Latin adjectives used substantively like rivāl-is and tribūnal many of which have been adopted to English as well (*rival, tribunal*), either directly or through French.

A closer look at the *–al*-adjective/*-al(s)*-noun candidates reveals again a very varied picture. Many long-established ones have entries as adjectives and as nouns. Others which seem to be established for equally long amounts of time like *chemical(s)* only have an entry for an adjective which nevertheless comprises sample sentences of nominal use of *chemical. Journal* is traced back to the French language where the word-combinations *papier journal/livre journal/registre journal* got ellipted to *journal.*

Again we have examples which are commented in the OED as being an "elliptical use" of the adjective as in the case of *horizontal* from *horizontal line, bar, member* and words whose meanings are not mentioned in the OED at all in their nominal use like *physicals* for *physical properties* and *lineals* for the *side parts of doors and windows.* Here at least, an adjective-noun conversion has taken place as in the well-known example *multinationals* from *multinational companies* – which can also be found in the corpus.

In contrast to the above-mentioned *–ant/-ent* examples, there has always been an equivalent *–al*-adjective for the established *–al(s)*-nouns.

Adjectives in *–ic* – Nouns in *–ic(s)*

Apart from sharing the Latin origin and equivalents in the French language with the preceding examples *–ics*-nouns are also products of an analogy process as there is a long tradition of naming systems of thought and knowledge, hence sciences by the help of the ending *–ics* (cf. OED-entry for –ic). The corpus examples of *analytics, hydraulics* and *kinetics* account for this phenomenon. Some examples have an established entry in this field like *optics* (the theory of the laws of sight, cf. OED-entry for optic) and *aesthetics* (the criticism of philosophy of taste) but acquire a different meaning in the domain-specific use: *optics* stands for *fiber optic lines* or *cables* and *aesthetics* for the *aesthetic quality* (source: ENGICOR).

And there are more examples which back up the idea of adjective-noun conversion having taken place like *aromatics* from *aromatic compounds*, *aromatic hydrocarbons, inorganics* from *inorganic compounds*, *contaminants* and *material(s)* as well as *caustics* from *caustic soda* or *caustic solution*.

There is one example where there was no established *–ic*-adjective first: *cybernetics* was created by analogy and *cybernetic* then produced by backformation (cf. OED-entry for cybernetic(s)).

Recapping

Summing up, it can be said that looking at adjectives ending in *–able, –ive, -ant/ent, -al* and *-ic* and their corresponding nominal use, certain similarities can be displayed but it must be stressed too that it is far from being a homogeneous process.

Leaving adjectives in *–ous* aside, it can be deduced that evidence was found in the corpus and backed up by the OED-dictionary entries for the existence of an adjective-to-noun conversion process where adjectives of the different endings are active to a certain and apparently different degree. The difference in degree is influenced by the etymology of the words where all adjective endings have in common that they started their generic lives in the Latin language from where it was directly filtered into English or via the French language. It also became clear from the examples, that not only did the adjectives in question originate in Latin and continue to develop in French but that also the process of adjective-noun conversion started at least for some of the endings in Latin and occurred in French as well. Furthermore, it has become obvious that this is a process which goes on even nowadays and that there were examples established which can be regarded to be at different stages of the process.

From some of the examples, it can be inferred that other processes such as analogy and derivation - in the case of the suffix *–ant* added to verbs directly to form nouns without employing adjective-noun collocates first – are active at the same time and produce very similar results, so that every single instance needs to be looked at individually.

Still, having now established that the adjective-noun conversion can happen to different sorts of adjectives, the question is to what extent this happens in language use.

To these ends, prospective *–ables*-nouns being hypothetically converted from *–able*-adjectives were searched for on the Internet. ENGICOR has 365 adjectives endings in *–able* which were entered in their *–ables*-noun form into the Google search-engine. Far from being a dull search exercise, the investigation brought interesting facts to light. From the 365 tested items only 17 (4.65%) did not have any entry on the internet. As Google does not lend itself to giving information on reliable numbers (cf. corpus list discussion, http://listserv. linguistlist.org/cgi-bin/wa?A2=ind0503&L=corpora&P=R6197#TOP), the exact number of occurrences were just regarded as informative though not of having absolute relevance.

The scale went from just one entry as a kind of nonce-formation (*baleables, sorbables, non-permeables*) and items that occurred less then ten times (*autoclavables, divertibles, impenetrables, inextensibles, metabolizables, remediables, swimmables*) to items which occur several hundred thousand times and thus being very much established (*deliverables, flammables, inflatables, payables, reproducibles*) (as of March 2005). Google-searches produce data which contain a lot of noise (unwanted and disturbing information) but trends were still clearly visible. Apart from belonging to domain-specific texts in the field of science and technology, the examples often named groups of people (i.e. of pressure groups and in band names), and names of goods and products (brand names). Noise came in the form of numerous French and Spanish texts (although the search language was set to English), and English texts written for instance by Spanish native speakers and containing translation mistakes (as the adjectives inflect for numbers in Spanish leading to the same forms).

The investigation was thus so surprising because many findings counteracted the intuition of the searcher. It is beyond the scope of this study to look at all examples; still, having had a closer look at some they all made perfect sense in their specific context (*fishables, regrettables, teachables*).

In short it can be claimed from this search that every adjective ending in *–able* is in practice able to produce a noun ending in *–ables* either by ellipting the noun from a well-established adjective-noun collocation or directly naming an object by taking an adjective which describes an outstanding characteristic of the named object and create the corresponding *–ables*-noun.

Having looked only at words ending in *–ables* it can be questioned whether the adjective>noun conversion displays all the characteristics of a full conversion, i.e. to behave inflectionally like a noun and consequently vary in terms of number and case. To this end, the corpus was searched to see if the plural forms in –*ables* have any singular equivalents. In fact, only one item was found in the singular form: *a deliverable* (as in: *This deliverable includes field instrumentation, IT infrastructure, and software for data acquisition, data analysis and data representation.*). *Deliverable* occurred four times in its singular form in the corpus. Again the search was extended to the internet and the established *–ables* forms were looked for in the singular with the help of the Google search engine. The found data again reflects the process character of the phenomenon. Some items occurred very frequently on the web and seemed to be well-established like *an imponderable* and *a deliverable.* Others occur more infrequently, often being part of an explanation as in the case of *a recyclable* or *a refillable* (as in: *What is a recyclable?*). Others again are only mentioned once as *a non-renewable* or *a biodegradable* and a few did not occur at all. It has to be stated that the plural use of the *–ables* words is much more common and that the singular use of only a few *–able* words as nouns are well-established (cf. OED entry of *imponderable* – first established use in 1842), whereas others are still about to enter the language use and others may come into being in the future depending on whether the need will exist for a new singular or not.

Another feature which was observed when looking at the *–ables* words is the fact that 6 out of 19 items happen to be in their negative form (*non-durables, non-recyclables, non-renewables, non-refillables, non-usables, unrecyclables*). It was tested whether the corpus would provide the same amount of *non-*ives, non-*ics, non-*als,* and *non-*ants/ents.* And apart from *nonmetallics, nonrespondents* and *non-participants* it did not. But when negative adjectives with these endings were searched in their plural noun form on the internet, it revealed that these forms exist as well (*non-explosives, non-invasives, inactives, ineffectives, noncarcinogenics, nonionics, non-biologicals, non-residuals, nonexistents, non-permanents*). So the negative forms produced by the help of *non-* is no privilege of the *–ables* forms but reflects a characteristic of the world in itself: the opposites of objects and entities can be described by negatives of the positive term.

The Semantics of the Resulting Language Items

With view to teaching, it is worth looking at the meanings of the products of adjective>noun conversion processes. In this regard, it is important to highlight those developments in language where changes have occurred recently and where they can be expected in the future.

Our class of *–able*-words shows the clear tendency to denote a class of items, materials or goods which are x, if x equals *durable, imponderable*, etc. Some of them fulfil a passive function like *renewables, consumables, extractables* and *deliverables*, as they can be described as something which can (not) be x-ed, if x equals *renew, consume, extracd* and *deliver*. One example *float* has an active meaning and can be expressed by something which can x, if x equals *float*. Two language items, i.e. *durable* and *imponderable*, due to their origin in the French language and the lack of corresponding English verbs can only be expressed by something that is x, if x is *durable* and *imponderable*. The range of items found in the corpus displays a few long-established examples like *moveables, consumables* and *imponderables* but it reflects the tendency to produce new words in the field of environmental engineering English to which *biodegradables, recyclables, renewables* clearly belong. Their relative newness in language can also be claimed on grounds that most of them have not yet found entry into the OED (*biodegradables, recyclables, refillables*).

Concerning words ending in *–ics*, two classes of words are distinguishable based on their meaning. There is first a group of words which denotes chemical substances like *aliphatics, aromatics, caustics, inorganics* and *phenolics, styrenics*, where it is likely that there used to be a second element, a noun, that has been ellipted (i.e. *aromatic compounds, caustic soda/potash, phenolic plastic, styrenic polymers*). Some non-chemical words might have gone the same way like *geosynthetics* from *geosynthetic material* and *toxics* from *toxic substances*.

And secondly, there is the group of words denoting sciences, branches of sciences and theories which are built according to their examples from classic times like *ethics* and *rhetorics*. *Hydraulics, dynamics, kinetics, mechanics* and *physics* are examples of this kind. Here, it was not an ellipting process that took place; the analogy to the classic forbearers must have been the driving force behind the production of these words.

Looking at words like *abrasives, additives, adhesives, corrosives* and *intrusives*, it can be assumed that again a substance-like word that constituted the second part of a two-word compound must have been deleted along the way and the result was a word denoting an x substance, if x equals *abrasive, additive, adhesive, corrosive* and *intrusive*. Whereas the mentioned words clearly belong to the field of chemistry, words belonging to more unspecified fields have functioned in the same way: *cooperatives* from *cooperative shops* and *farms*, *objectives* from *objective point* and *positives* from combinations like *positive aspects*.

And the same development can be observed with words ending in *–ants/ents*, where the majority of the found words denoted a chemical agent and therefore belonged to the field of environmental engineering English, like *adsorbants, precipitants, reactants, coolants, solidificants, flocculants, leachants, lixiviants*, etc. Interestingly, their definition in the OED very often exists in the form of "an x-ing agent", and that is also the form in which still a lot can be found in the corpus: *leaching agents –leachants, coloring agents –colorants, reducing agents – reductants*. This can be regarded as a sign of a very recent process and it is claimed here that we can expect more of this development in the future of environmental engineering English. Having found some examples of *foaming agent* and *bulking agent* in the corpus, the internet was checked to find out if these terms exist in the form of *foamants* and *bulkants* (because there were no instances of *foamants* and *bulkants* in the corpus). It was found that whereas *foamants* occurred already about 200 times on the internet, *bulkant* did not. Still it can be regarded as a possible future word. Further evidence for the productivity of *–ants/-ents* and their established meaning of "agent" comes from another process which results in the same word format. *Surfactant* is a blend of *surface* and *active* + the ending *–ant* and denotes a substance called *surface active agent* in its long form. So the use of *–ant/ent* to denote an agent substance is applied in spontaneous word formation as well as in conscious word manufacturing.

Other examples of words ending in *–ants/ents* denote personal agents such as *inhabitants, occupants*, and *participants*, but are well-established and common in General English as well and therefore not expected to yield new developments.

With the class of words ending in *–als*, clear common classes of items could not be established. Words like *annuals, perennials* and *periodicals* share at least their reference to time; words like *horizontals, laterals, lineals, peripherals*, and *terminals* retain their reference to a certain position, and words like *locals, officials, individuals* and *professionals* denote kinds of persons. Still, examples were not so numerous and recent developments were only established in the cases of *horizontals, laterals*, etc. and there was one example where *physicals* denotes *physical parameters*.

Summing up, it can be said that environmental engineering English is an area in which new words get formed. This happens above all in the field of chemical substances (*-ics, -ives* and *–ants/ents*) and classes of items/materials and goods (*-ables*). By the sheer amount of processes, the words in *–ants/ents* and *–ics* were the most numerous, words ending in *–ics* and *–ives* to a lesser extent. It is assumed here that all the found data reflect the movement towards a more analytic language where the meaning of the words is clear from the context and in a further step the words acquire the status of "specialist words" in the field of environmental engineering and its neighbouring specialist sciences and technologies.

Conclusion and Outlook

From this case study, it has become clear that far more research on the semantics as well as on the extent of use of adjective-noun conversions and similar processes has to be carried out in order to gain the full picture of the development. The present study made but a start. The investigative means in terms of corpus suitability and the allocated time were restricted.

Still, the retrieved data can be regarded as showing the clear tendency of adjective-noun conversion to occur more often than has up to now been observed in more language domains than thought to date.

Dictionaries as well as standard works on word-formation and grammar do not provide a full account of the reality and learners might have problems in establishing the meanings of searched words with the help of reference books. The word *fines* in the sense of fine particles can serve as an illustration for this problem because it is only under the entry *fine* (adjective) where one could find sample sentences which reveal the meaning of *fines* and not under the entry *fine* (noun) where a learner would be likely to start his/her search.

The shortcomings could be remedied by extensive corpus studies, preferably with a large diachronic corpus which could be investigated in time divisions in order to document first the established adjective-noun collocations and then to observe the change in quality towards a converted deadjectival noun.

The examples found in the corpus revealed on the one hand a clear tendency to belong to domain-specific language (*recyclables, biodegradables, refillables* as specific for environmental engineering English) but on the other hand, further research showed that this can happen in all language domains as well as in General English (*personables, unbearables, capables*). Linguistic literature has not yet taken account of the versatility of the process. In this study, it is claimed that new nouns displaying original adjective-endings like *–ive(s), -ant(s)/ent(s), -al(s), -ic(s)* and *–able(s)* are evidence for the general and wide-spread possibility of an adjective-noun conversion process where it remains questionable to what extent the elliption of a single noun takes place as opposed to a class displaying a common semantic preference being ellipted in other cases. It remains to be investigated too to what extent classes of objects can be named directly by the adjective forms used as nouns denoting an outstanding characteristic of the concerned object.

It was the occurrence of words ending in *–ables* that commanded attention in the first place. But when looking at the simple frequency indicator another ending is more frequent. Among all established derived forms from the first 500 keywords, there are 12 forms in *–ants/ents* (like *solidificants, toxicants, leachants, colorants, extractants,* etc.), six forms in *–ics* (like *analytics, site-specifics, organics*, etc.) and only three forms in *–ables* (*recyclables, biodegradables,* and *extractables*), whereas it remains questionable whether all forms were formed by conversion or whether analogy could have played a role as well.

Finally, it has to be said that learners of domain-specific language variations should be made aware of the process as it is more frequent than studies have shown so far. The look back at the adjective-noun collocations might be a means of making the meaning of the converted forms clearer to the learner. In most cases, typical contexts as for instance displayed in concordance lines should be enough to understand their meaning in use. Concordance searches of the related adjective can turn the learner into an investigator on the search for the implied meanings.

5.2.10 Others

5.2.10.1 Word Manufacture

Word manufacturing is a very rare process. Examples for this process cannot be systematically detected by concordancing; so the only two examples from ENGICOR were found by chance. According to Bauer who says that the "purest cases of word manufacture are when a word is created ex nihilo, with no morphological, phonological or orthographic motivation whatsoever." (1983:239), ENGICOR provides one purer and one not so pure example of word manufacturing.

The first example *gunk*, was first used as the trade name for a detergent solvent before it was used informally to name any kind of viscous or liquid substances (source: OED entry for gunk). Here word manufacturing took place without any existing word material just as a completely new creation, a phenomenon that occurs often when naming "new scientific discoveries, trade names" and "names of organizations" (Marchand 1969:452). *Gunk* occurs 7 times in the corpus, all in the same text (3 times in inverted commas, four without). The author using this word even specifies its meaning as "*the material removed by the centrifuge*", so he does not assume it to be known to the readers in this specific meaning.

For the second example – *biosolids* - already existing word material was put together. According to its form it is a noun resulting from the combination of a neoclassical combining form and a noun but a corpus quotation as source of some anecdotal information proves that there was a naming competition to find a new name for sewage sludge, which had at the time acquired a bad reputation:

> In their ironic 1985 book Toxic Sludge is Good for You! John Stauber and Sheldon Rampton recount the amusing story of how the Water Environment Federation (the WEAO‘s U.S. cousin) struck a name-change task force to determine a PR-friendly name for sludge, the spreading of which on farm fields had come under attack. In June 1991 the task force selected the term "biosolids" from more than 250 suggestions that included "humanure," "geoslime" and "bioslurp."

(source: http://list.web.net/archives/sludgewatch-l/2002-February.txt , accessed on 09 August 2005)

5.2.10.2 Loan Words

Again, candidates for this category cannot be searched for systematically in the corpus but *schmutzdeck* and *trommel* were detected by chance to have found their way from the German to the English language probably following the transfer of a technology. Loan words are words which were adopted from other languages and often travel together with the new concepts or objects that designates them. *Trommel,* which is a rather general term in German (i.e. drum), in English acquired a rather specific meaning in the field of mining where it is a rotating cylindrical sieve used for washing and sizing ores (source: OED-entry for trommel). In ENGICOR it occurs in texts on mining as well as waste treatment and recycling which shows that it has travelled even further through the language (ENGICOR: *A trommel is a cylindrical drum screener that conveys material.*). Once in the English language, it started to form derivations. The corpus contains instances of the verb *trommel* (*Once this waste was trommeled,..*), of *trommeling* as a noun naming the process and the adjectives *trommelable* and *untrommeled.*

Schmutzdeck or correctly spelled *schmutzdecke* is a biologically active filter layer which forms on the surface of slow sand filters and removes suspended organic material from contaminated water. Its hyponym is *biofilm* but *schmutzdecke* replaces *biofilm* in the specific context of slow sand filters. Its origin remains opaque because it cannot be proven that it travelled together with the technology transfer: in combination with filter, the Google-search finds 391 English but only 6 German websites (accessed on 09 August 2005). Furthermore, the clumsy German equivalent to *slow sand filtration* - Langsamsandfiltration - gives reason for speculations that this technology was developed in an English speaking country first.

The word *pitchblende* is an example of a mixed loan translation where the first element was translated (German Pech > English pitch) and the second element kept its German original form. It is just one example for a few words in the field of mineralogy which found their way into the English language (i.e. *feldspar, hornfels,* etc.).

5.2.11 Semantic Change

Semantic Change is by its very nature not a word-formation process because the words in their specific form already exist. It is the meaning of the words that changes when words from General English or other domain-specific languages are used in a particular language field in a new sense.

Moreover, results from semantic change cannot be systematically searched for in a non-annotated corpus like ENGICOR; the following examples used for illustration purposes are all findings arrived at by chance. Semantic change, however, is a means to extend the vocabulary of a specific language and is therefore looked at briefly here, as students need to be aware of this kind of language development as well.

Resistance is a very old and established example of semantic change. Stemming from the idea of the ability or willingness of people to withstand enemies or difficult situations its meaning extended to objects later on to oppose any kind of material force like pressure or movement and even later the more specific property to withstand electrical current.

And it is also in more recent examples of semantic change that a particular property of the original word was the motivation for using it in the naming process in environmental engineering English. In *scrubber* for a big cleansing apparatus, it is the meaning "device that cleans" which must have been the foremost idea in the naming process, and in *cyclone* for a machine that uses centrifugal force to separate out solids it is the similarity to the storm that motivated the naming process. In *grizzly* for a *size separator* it might have been an even further removed characteristic ; the mightiness of the machine that can be compared to the huge animal.

Although results from semantic change cannot systematically be searched for in the corpus, it can be proven by corpus means that there has indeed been a semantic change. If for instance *pregnant* is taken in its specific sense in ENGICOR (describing a solution which is rich in dissolved metals or minerals leached from ore) and compared to its general meaning, it is obvious that we face two different lexical items, committing themselves to totally different collocational and colligational patterns as well as having different semantic preferences and prosodies. For that purpose compare Table 34 (usage profile of *pregnant* in ENGICOR) and Table 35 (usage profile of *pregnant* in a subcorpus of the BNC, in the register of news, accessed via Mark Davies' BCN interface VIEW). In ENGICOR *pregnant* is used to modify a solution whereas in General English it is only ever used to refer to females expecting offspring. In rarer cases it can also mean "heavy with" (e.g. hope). No verbal collocations are attested in ENGICOR and neither are there any colligational patterns like *pregnant with*.

Tab. 32: usage profile of *pregnant* in ENGICOR

pregnant adjective 114 occurrences in ENGICOR	
N-1	**N+1**
	pregnant liquor (8)
	pregnant lixiviant (6)
	pregnant solution(s) (38)
clusters	
pregnant leach solution (33)	
pregnant solution pond(s) (6)	
pregnant elute/leach/stripping liquor (11)	

Tab. 33: usage profile of *pregnant* in BNC, subcorpus register of news

pregnant adjective 114 random samples from the BNC/News	
N-1	**N+1**
BECOME pregnant (14)	pregnant woman/women (20)
GET pregnant (5)	
BE pregnant (20)	
clusters	
four/five/six/… month pregnant (7)	
pregnant with (7) William/her second child/…	
pregnant with (5) hope/the future/initiative	

Moreover, the occurrence of the opposite *barren solution* in ENGICOR shows how successful the use of language similes of this kind are, as the term *barren solution* designates a solution which has chemically been stripped of its metal and mineral components and in this way serves as the opposite of *pregnant solution*.

5.2.12 Summing up

The foremost materialisation of a language is its lexis. Vocabulary learning is what learners associate language learning most with. A lexical approach to language learning implies a way of making the learners familiar with the typical language use as well as with additional information on the structure of lexical items. This is when morphology comes into play. Unlike native speakers who rather have an implicit knowledge of the means of word-formation in a language, I claim that learners of a domain-specific language need to be made aware of the different kinds of information hidden in the different derivational patterns and all the other word-formation processes. With view to minimising the amount of effort required to learn and a useful categorisation of the lexical items, the relations of the words within their morphological patterns, the meaning components added by affixation or combining forms, and their added syntactic information help learners to understand text or utterances better and produce their own language. Further to understanding the meaning of lexical items from their situational and textual context, understanding of morphological information is vital for the language learning process.

In a specific language, lexis serves as a primary means to express scientific concepts in a short and precise way. Specialists exchange information on processes and technologies and describe results from investigations and studies. Language economy is achieved by morphological means. Complex noun- or adjective-forming compounds as well as concise derivations are just an example for giving as much information as possible in the shortest possible way. They are the realisations of the prevailing motivation behind word-formation processes in the domain-specific language: the labelling of new concepts.

Bauer quotes Bloomfield as follows: "the process of linguistic change has never been directly observed" (Bloomfield in Bauer 1997:12) and further on claims that "lexical change is by its very nature unsystematic" (1997:29). ENGICOR as a means of making observations possible as well as tracing some lines of if not systematic but at least likely change is not an ideal instrument, but the amount and quality of findings it revealed permits useful insights into the morphological peculiarities of environmental engineering English and produced extremely valuable linguistic data for material design.

When embarking on the challenge to learn a domain-specific language, learners do not need to learn a whole new system of morphological structures. Every means of word-formation examined can be found in General English as well. It is rather the difference in the extent of use of certain word-formation processes that differentiates the domain-specific vocabulary from the general. Environmental engineering English like any other specific language favours some ways of forming new language items over others and learners benefit a great deal from knowing where this language change takes place and what the resulting meaning of the new elements is.

Coming back to the syllabus design, not every word-formation process can be of the same importance. It is certainly compounding and derivational processes that should be concentrated on in detail as well as the use of combining forms and conversion which was described above as a prominent feature of a domain-specific language. No claims can be made about the importance of processes of semantic change as the corpus was not suitable for this kind of linguistic information, but blending, clipping, backformation, word manufacturing and the introduction of loan words can be considered to be of minor importance.

Students should be made familiar with the sort of elements that combine or undergo change in the different word-formation processes and with the form of the results as well as of the meaning load added. As texts which can be defined as academic prose in a certain field are characterised by a huge amount of nouns and noun phrases, the morphology of nouns and their immediate co-text should be central to the teaching process. From the investigation of the lexical profiles of the first 500 keywords in environmental engineering English, the central role of noun compounds as well as modifying adjectives became clear. In practice, familiarising learners with morphological information could include the following: identification in text, discussing meanings loads, paraphrasing the shorter items with self-produced explanatory sentences and use in own language production. The format of exercises can be manifold: *a wind-driven device* can be described by students as *a device driven by the power of the wind*, *pollutant-free* can be changed into *a state where all pollutants are removed*. The same can be applied in reverse where explanations are given and students have to find the specific term.

The semantics of environmental engineering English-specific terms can be combined with the specialist knowledge of the learners from their specific engineering background. Examples of compounds can be searched for in order to be categorised as belonging either to the category of processes/operations, process indicators, process agents or products, process equipment or process locations and afterwards be used to describe a whole technology. The development of combining elements such as *phyto-* can be explained in parallel with the emerging of the new technologies which use plants to remediate contaminated soil and groundwater and in this way exemplifies how language change is coupled with technological progress.

The are also other ways of turning the language learner into an active participant in the learning process and into a researcher. With the corpus-information as background knowledge, the teacher can devise search tasks for the students to use the corpus again. Where do the conversion processes described in the *–ables* case-study originate? What kind of change did the word undergo? I can say *fuel-*

rich and *clay-rich* but would I also say *pollutant-rich*? These are only a few questions that can make students aware of conversion processes, processes of semantic change and series-forming compounding elements and let them devise their own answers. Students will be more independent in future language learning situations also beyond the actual classroom situation.

It can be seen from these examples that acquiring knowledge of word-formation patterns is not detached from using specific language in communication. This is what should be aimed at, combining language learning strategies both of the collocational and morphological kind in order to make students realise the information potential of lexical items. The implicit morphological knowledge of the native speaker can thus at least in parts be compensated for and language learning processes shortened: recognising the meaning load added by a suffix or being able to understand the implied information of an adjective-noun conversion reduces time spent on language learning and effort.

6 Chapter 5 - The Application of Corpus Data in Language Teaching

6.1 The Aims

The need for engineering specialists who can master a foreign language to discuss subject-related topics is growing on a parallel with technological and scientific progress as well as co-operation on an international scale.

When entering Universities of Applied Sciences, students usually have a good grounding in General English and that is where they start their journey which aims at enabling them to communicate in professional situations and deal with the specific topics in the relevant oral and written text types. The way this journey takes needs to be carefully prepared by the ESP-teacher.

From what has been discussed in earlier chapters (cf. 2.1.4 and 2.1.5), there is strong evidence that language comes in chunks and that meaning is spread over multiword items rather than being presented in single words. In order to become a proficient user of a (domain-specific) language, the students need be enabled to use multiword items in their utterances and not translate word for word. Mukherjee calls this "language routines" which are a prerequisite for native-like fluency (cf. Mukherjee 2002:78 ff.).

Sinclair (1996b:114) claims that "A word becomes associated with a meaning through its repeated occurrence in similar contexts". This process runs through language whenever a new word comes into existence or a new meaning of an existing word appears, but at the same time every user of the language repeats this process in his or her individual learning process (cf. Hoey 2005). Learners of a second language do not have such a vast exposure to language material as native speakers have when learning their language during their lifetimes. Therefore the language material they are exposed to should be carefully selected.

This selection process is crucial and a lot of work goes into the research for the suitable material, discussions with subject specialist teaching staff and target situation analyses. The claim of this thesis though is that corpus analysis adds a surplus of information on the topical as well as on the linguistic level in order to ensure the highest quality in the selection process.

The attempts to make sure that only relevant language material is being taught are manifold and I just want to give a reminder of West's General Service List (West 1953) and Coxhead's Academic Word List (Coxhead 2000) mentioned earlier. And to say it with Coxhead's words, what needs to be taught should be based on "principled decisions", which in turn need to be based on empirical data. Within the framework of this study, I went for the list of the first 500 keywords stemming from a comparison of ENGICOR with a large general corpus, the BNC. As this is just the first step in the selection process, the first 500 words were analysed in their linguistic context and the findings documented in the usage profiles (see Appendix B). The keyword list can be compared to a backbone to which the flesh in form of the collocational information needs to be added. This is how the researcher/teacher gets to a kind of reference material, in my study the 500 usage profiles of the investigated keywords which can be regarded as a extended micro-glossary and a micro-grammar in the Firthian sense mentioned earlier.

These usage profiles are a kind of knowledge bank which can serve as a basis for teaching and learning a domain-specific language. Having now on the one hand the need to enable students to converse fluently in this language and on the other the wealth of the worked-on corpus information the question is what can be done to make use of the data in order to achieve the teaching goals.

The aim is to make use of the corpus information in the pre-, in- and post classroom activities. This reflects the different tasks of ESP-teachers at university level. They need to decide which topics to teach, decide on which language items to focus on, decide on suitable materials and adequate activities in the classroom as well as beyond when focussing on autonomous learning activities and preparing the ground for lifelong learning. So the questions that need to be answered in this chapter are (1) How can corpus data help in syllabus design?, (2) Can I use corpus information as well in materials design?, and (3) In what way can it be applied in classroom situations and beyond in the process of acquiring strategies for lifelong learning?

Before I go on to look at these questions in detail, I want to examine the rationale behind the necessity of focussing on corpus data and subsequently phrasal language material in the following paragraph.

6.2 The Rationale

Drawing on insights from early corpus research, as early as in 1988 Sinclair and Renouf described the model of a lexical syllabus for language learning which should comprise "the commonest words forms, their central patterns of usage and the combinations which they usually form" (Sinclair/Renouf 1988:148). Later these ideas were developed further into the lexical approach by Lewis (1993, 1997, 2000) and Willis (1990) who argues for a strong emphasis on teaching lexical chunks and intense exposure to the commonest patterns and meanings in the language (Lewis 1993). Apart from this focus on the importance of prefabricated language, we can also find arguments for the concentration on awareness-raising exercises in order to show students how language works and how it is best learned in Lewis (2000:Chapter 8).

As a matter of fact, many studies (Bowker/Pearson 2002, Bahns 1993, Lawson 1998) have proven that students "are more likely to let themselves down by using the incorrect collocates than by using incorrect terms" (Bowker/Pearson 2002:155). Bahns and Eldaw for instance found that "although learners can comprehend and translate English sentences containing collocations (Bahns/Eldaw 1993:108) there is no way to "easily paraphrase your way around collocations" (Bahns/Eldaw 1993:101. As a reason for that phenomenon Lawson argues that "collocational behaviour is carried over from the first language" (Lawson 1998:227). However, the central role collocation plays in language use cannot be overlooked as collocation is, according to Clear, "a powerful organising principle in English" (1994:33) which let many linguists come to the conclusion that knowing a word means more than knowing a one-to-one translational equivalent of a word in the mother tongue. It means also getting to know its collocational and colligational patterns as well as its behaviour in terms of semantic preference and semantic prosody.

There is no way to avoid learning chunks of prefabricated language and for that purpose we need to know which chunks are typical, that is most frequent in our domain-specific language in question as we have seen that every ESP-variety has its own usual word combinations which are used again and again by the speakers of discourse community (cf. Williams 2002). And it is knowing and being able to use these typical collocations that turn a language learner into a proficient user of a specialist language. According to Pawley and Syder (cf. 1983), speakers of a language possess a large number of "lexicalised sentence stems" which are stored in their minds. Putting together words according to the open choice principle demands much effort from both, the speaker and the listener. So we have apparently a large memory storage for collocations whereas our processing capacity for information broken down into single words is limited and not working very efficiently.

The consequence of that can only be to concentrate to a large extent on the teaching of vocabulary in its typical phraseology to the students. Bahns sees their number as the "main obstacles for teaching collocations systematically" (Bahns 1993:56). Moreover, it is also their unsystematicity as paragraph 4.3.2 has shown us. What the teacher can do to pave the way to incorporating collocation into language teaching is the subject of the following overview on the use of corpus data in language teaching.

6.3 Making Use of Corpus Data

The line of argumentation is now about to getting closer to tying up its loose ends. I have worked my way through the selection of the most relevant linguistic material and its analysis on a collocational and a morphological level with different sidelines concentrating on validity and relevance of the investigated data for the described target group: students of environmental engineering studying English for their specific professional purposes. Now it has to be demonstrated how data from the empirical analysis is applied in the different processes of teaching and learning a specialist language.

Looking at this problem first only from the teaching side, the main questions are what to teach - that is which syllabus decisions are to be taken - and second how to teach it - which aims at the applied methodology. Both questions have been answered earlier in this study in short: the 500 most frequent keywords and their respective usage profiles is what should be concentrated on in the classroom and the fact that the teaching process has to provide multiple encounters with these keywords and their linguistic environment is in short how this has to be organised. This simplified view is now going to be elaborated.

6.3.1 Pre-Classroom Use

6.3.1.1 The Syllabus

Before any classroom activities can be started there need to be considerations concerning language material the course should concentrate on; a syllabus has to be developed which functions as an organising principle for the course and the teaching processes. For that purpose I have to go back to the 500 keywords and establish the order in which they should be dealt with.

First of all, these 500 keywords are already a selection from the whole of the words featuring in ENGICOR. On a more refined level, I took a look at which words were more frequent in one of the subcorpora compared to the whole corpus (compare 4.1.4./see Appendix A). The information on this level helps to get an idea which words should be the starting points when embarking on special superordinate topics such as mining, remediation techniques, recycling techniques, etc. As they are the most common words when dealing with these topics, they should get attention first in the teaching process. At the same time, words which occur more evenly throughout the whole corpus should also be focussed on (i.e. *amount, quality, effective, reduce*, etc.). This means that they should be dealt with in a way that highlights both their general use which spans all over the corpus and their specific use in the different subcorpora (i.e. the universal concepts of *waste generation* or *waste disposal* vs. very specific concepts like *waste rock* from the mining subcorpus *or packaging waste* from the waste and recycling subcorpus).

Having dealt with words which are the most topic-specific, I can now move on to the next level which is represented in the form of the keywords of a specific text. This procedure was exemplified in 4.4 which showed that the keywords of such a text compared to the whole corpus reveal which words are foremost used to express the content of the text and which consequently would be needed to reproduce its content again by the students.

From there we can move on to the most refined level which is the word level where teaching concentrates on the single keywords in their linguistic context. The reference material for that are the usage profiles. Here the information on collocational and colligational behaviour as well as on the patterns of semantic preference and semantic prosody which the students need to acquire during the learning process in order to efficiently use a linguistic item in speech are to be found.

Speaking at this point of "words" relates to the form in which the linguistic items are found in the different lists and it is understood that their usage in multiword items is implied at the same time. There has been discussion on whether there should be a shift of emphasis from word frequency lists to "phrase frequency lists" (cf. Stubbs 2002:215) but as phrase frequency lists would include a vast amount of data and generate problems related to the different structures and forms of the included phrases, another solution for providing and displaying phrasal information needs to be found, which is in the case of this study the usage profiles as a kind of inventory of linguistic information.

Syllabi are often graded according to difficulty and complexity of the linguistic items. I argue that difficulty and complexity are no longer suitable for grading or better to say structuring language material at advanced level courses at university. Taking their place topicalness comes into play again. A syllabus for a specialist language class has to respect students' needs, that is the topics they have to deal with in future professional situations and the way they have to deal with them. At this point I need to resort again to the results from the questionnaire sent to the companies which collected the expert opinions on the topics most dealt with in their jobs.

To illustrate the procedure of syllabus design and to sum it up, the two most superordinate topics in environmental engineering in our university course are on the one hand recycling of materials and on the other remediation of soil and (waste)water. These two topics are thus represented in the two subcorpora remediation technology and waste and recycling. A third topic, mining, is represented in its own subcorpus although it can be subordinated under remediation technologies. Looking at the subcorpus-specific words and their specific co-text when dealing with the respective topic is thus the first step (as well as keeping an eye on the words which evenly occur throughout the whole corpus). Then the specific text's keywords should be looked at before going into detail and looking at their immediate co-text.

Working on the syllabus structure is not as absolutely a linear process as it might appear in the model of levels for syllabus design. The work with information based on the usage profiles – i.e. the collocational information – will already occur when dealing with the subcorpus-specific data as students are not supposed to deal with isolated words but with multiword items.

It can be claimed that the process of syllabus design is entirely corpus-driven as no preconceived assumptions play a role. The items with the highest frequency deserve the highest priority within their topic.

6.3.1.2 Materials Design

For the materials design as well the corpus data in the form of the keyword list on corpus, subcorpus and text level and the information from the usage profiles need to be used as reference material. The principled decisions taken during syllabus design have to be consequently taken into account when designing teaching and learning material. This implies that the material chosen has to ensure the multiple encounters we need in order to achieve priming as well as to serve as samples of exemplification in case of conscious instruction in vocabulary teaching.

Furthermore, it needs to be mentioned that all the other common features of modern material design apply as well, as for instance the need to use authentic and motivating material. It is of paramount importance that the pieces of language chosen for exemplification or practice purposes are taken from the future professional context of the students or at least prepare the ground for it.

One corpus-based example for ground-preparing material which can be used in intensive vocabulary work will be introduced in the following paragraph.

6.3.1.3 Recontextualisation - Using the Information from the Usage Profiles

Having established the format of the usage profile (compare 4.2) of the found data, the next question is how this can be of use for the teacher in his or her efforts to design valid teaching material as well as in extracting all the condensed information again for use by the students.

First of all, the usage profile as it is can serve as part of the teacher's general body of reference and information material with all its different levels of information. Before I go into detail when commenting on the use of corpus data in the conclusion, I will show one application of these data: I call this part of the procedure extraction of information. For an extraction text, I use the WordSmith 4 concordance tool again to show sample sentences that comprise the desired multiword items and try to put sample sentences in a certain order in such an extraction text. As an example see the following extraction of information for the *word infiltration*:

Fig. 12: Extraction of Information – Infiltration

Infiltration is the absorption and downward movement of a liquid into the soil layer or other media or vessels. *Infiltration* can either be desirable or undesirable. Accordingly, we have devices like drainage systems or capping and sealing methods to *prevent infiltration* or swales, which can be designed to trap particulate pollutants, to *promote infiltration*. Covers have also proved to be effective in *preventing infiltration* through mine tailings. Low permeability material like compacted soil may not be able to *prevent infiltration* entirely but is used to *reduce infiltration* to a certain amount so that the potential for seepage and leachate generation in landfills may be lessened.

The *infiltrating liquid* may for instance be water in the form of groundwater or precipitation. *Water infiltration* occurs through unsaturated porous material. Sometimes *groundwater infiltration into* underground mines needs *to be prevented* in order to avoid potential contamination of the aquifer. In other cases trees are planted to **prevent rainwater infiltration**. Again, *precipitation infiltration* might also be a positive phenomenon when *rainfall or snowmelt infiltration* is recharging the aquifer.

Precipitation and potential evaporation as well as properties of the soil material and/or other climatic factors influence *net infiltration. Net infiltration* has to be taken into account when developing a water balance for a certain site. High evaporation losses can result in *reduced rates of net infiltration.*

Infiltration rates are a function of a soil's moisture content, capillary pressure, unsaturated hydraulic conductivity, and the distance below the surface. There is no runoff or ponding when the *infiltration rate* is less than the saturated hydraulic conductivity. Runoff or ponding occurs when the *infiltration rate* is larger than the infiltration capacity and the saturated hydraulic conductivity.

In areas where there are *infiltration problems*, *infiltration devices* can be used to reduce discharge volumes and provide for groundwater recharge. *Infiltration galleries* for instance can be installed to enhance recharge. *Infiltration sumps* can reduce inflow in areas where the underlying soils are moderately to highly permeable, and the water is well below the ground surface. *Storm water infiltration pumps* are below-ground structures that are used to collect storm runoff and pass it into the soil.

Generally speaking, a liquid like water or any contaminant solution can *infiltrate* a medium like soil or waste rock, or a system, like a bioretention area, or a plume. When the *infiltrating liquid* contains greases or fats the clogging of infiltrative surfaces might be accelerated and no more infiltration occurs.

Water reuse can in some cases mean *reinfiltrating* treated water into a circulation pattern. The process is called reinjection or *re-infiltration*.

This is a text which I put together myself but which is strongly based on corpus examples. Almost all the sentences are direct quotes. There is only some minor cementing between some parts to achieve better cohesion. When taking a close look at these sentences, one can find huge jumps from one idea to the other, but I am convinced that students who have the subject expertise of environmental engineering can re-contextualise the loose sentences. In this way the loss of linguistic environment which occurred when the items are put in lists is reversed and the students see the words in their typical co-text, i.e. in multiword units as well as in their typical thematic environment.

In this format the most important multiword items and word formation patterns yielded by the corpus for one particular keyword are shown underlined. Moreover, it presents not only the relevant language knowledge but also, to a certain extent, the engineering context. Heringer (1999), who also uses a prototypical re-contextualisation method, speaks about putting flesh back on the bones with the help of corpus analysis (1999: 140). He argues that the sample sentences are without further context but not context-free. The context is re-established in the minds of the students, which leads to further associative thinking on their part, and being able to bring this about can be a challenging task for any teacher at the best of times.

The form in which the students are presented with the above information depends upon the methodological talent of the teacher (associogram, teacher presentation, gap texts, etc.).

It furthermore goes in line with findings from the field of learning theory that the richer the associations made with existing and newly acquired knowledge, the more readily-available the new pieces of language knowledge are.

In sum, recontextualisation of the information from the word list and the usage profiles is a method which should find its place in materials design and which stretches on to the use of corpus information in the classroom dealt with in the next paragraph.

6.3.2 In Class

I subdivided the in-classroom use of corpus data into two sections: priming and educational linguistics. This divide reflects two different routes language learning can take. They are on the one hand imitative learning when students copy language behaviour by starting to use prefabricated language because they have encountered it often enough to have stored form and all the different meaning relations in their minds. One the other hand, I put educational linguistics, a term by which I aim to cover all conscious reflection on lexis, be it entirely controlled by the teacher or even student-determined research.

6.3.2.1 Priming

Priming, as defined by Hoey means the following: "As a word is acquired through encounters with it in speech and writing, it becomes cumulatively loaded with the contexts and co-texts in which it is encountered, and our knowledge of it includes the fact that it co-occurs with certain other words in certain kinds of contexts" (2005:8). As a consequence it is the teacher's task to create opportunities for such repeated encounters. Once introduced into the classroom the relevant linguistic material needs to be recycled again and again in order to finally find its permanent place in the students' language competence. According to Hoey, priming "leads to a speaker unintentionally reproducing some aspect of the language" (2005:9), thus happening more unconsciously. This can be described as a communicative approach, teachers have to arrange teaching materials and activities in a way that allows students to encounter once introduced items over and over again.

6.3.2.2 Educational Linguistics

Educational linguistics in ESP-classes comprises two aspects: awareness-raising and explanatory instruction on language by the teacher and students exploring language themselves. It goes without saying that those two aspects are not an either/or matter but come in all sorts of combinations of the two. This can go from a traditional teacher chalk-and-talk activity developing a network of topic-specific vocabulary on the board over to a predetermined investigation done by the students but guided by the teacher to autonomous and self-determined serendipity-learning activities by the students themselves (cf. Bernardini 2000).

Instruction from the teacher's side can thus make intensive use of the language data provided by the corpus investigation. Mind maps are a good tool for introducing vocabulary to students. The following mind map of the search term "treatment" should serve as an illustration for this. Introduced to the student, this network of collocates can help students produce their own language in class and its spider-map form is thought to help associative learning and mnemonic reproduction at a later stage beyond the immediate classroom context.

Having such a powerful and reliable reference tool at their disposal is something that has never existed before and turns the ESP-teacher into a confident specialist of the specific language domain; it is no longer a case of years of prolonged contact and intensive work, but nowadays simply a matter of examining corpus data before starting the teaching process.

When I talk of educational linguistics I do not mean that engineering students need to become linguists themselves. But to a certain degree issues can be addressed at a level that is appropriate to learners of a domain-specific language which brings me now to the second aspect, the student-centred activities.

By searching for instance concordances students themselves can explore and detect the phrasal character of language. They can furthermore be confronted in their own searches with language material which shows them that words might have different meanings in General English and in environmental engineering English (see the example of *pregnant* in 5.2.11) or with certain word-formation patterns which make them reflect upon the productive character of language.

Students' own research activities can be geared towards different aims. It can be used for correction exercises, used to figure out certain meanings, can be form-oriented or used as an "hypothesis-testing device" (cf. Kettemann). It involves hypothesis-forming, searching and scanning, data-analysis and conclusion-drawing techniques and finally – in case of a successful learning process – the reproduction of the learned item in its correct collocational and colligational environment under consideration of its typical patterns of semantic preference and semantic prosody.

6.3.2.3 The Role of the Teacher

According to Danielsson and Mahlberg (2003) "the task of the teacher (is) to make the potential of natural language examples accessible to the students". As I have argued for a well-balanced mix of teaching and learning activities with an added plus of corpus use, teachers have to take on the corresponding further task and become the coordinator of the research and analytical processes in class.

The teachers have to develop a high degree of sensitivity towards the phrasal character of language and focus the learning process on it whenever applicable.

Having in mind which competencies are aimed at in the vocabulary acquisition process, it is important to offer suitable language activities in class which should involve observation and interpretation. These phases of reflection on language are where language learning at university level can assert its higher quality level. Teachers have to pass on important methodological skills like the ones discussed in the paragraph on educational linguistics which enable the students to become independent learners.

The biggest change in the work of ESP-teachers however happens in the field of the pre-classroom activities. Teachers no longer need to rely on ready-made materials which only in part fit the needs of the respective courses or struggle with the overabundance of data from the internet. The corpus as the provider of descriptive insights into the domain-specific language is a most powerful tool in the hand of well-informed and competent ESP-teachers, no matter whether they are native speakers or non-native speakers. Teachers thus add to their various roles the one of the applied linguist, a most challenging but at the same time very rewarding task.

6.3.2.4 Conclusions

Summing up, it must be retained that there are various methods to learn a language. Not a single one can guarantee success in language learning; it is rather a sound mix of them which makes sure various routes to lasting storage of prefabricated language in the minds of the students are addressed and thus motivates them. For that reason, teachers have to provide activities based on the different learning models which stretches from the above-mentioned imitative learning through repeated encounters to the traditional teacher instruction or student-centred discovery learning, in sum a course which combines communicative, awareness-raising and research-based elements.

Corpus-work as a relatively new element in the classroom aims at preparing students to become independent learners who can apply their acquired skills also beyond the classroom. According to Tognini-Bonelli, students, when dealing with corpora and undertaking concordancing, learn to understand patterns of language when they read vertically instead of horizontally (in a concordance). They also read fragmentarily instead of whole text (cf. Tognini-Bonelli 2001:3). This kind of fragmented reading can also be found when searching the web, so it gets close to modern information retrieval conventions. How students might apply their skills after class will be dealt with in the following paragraph.

6.3.3 Post-Classroom

I see the post-classroom use of corpus data and by now also corpus search techniques in three areas. There is first of all the almost well-established opportunity for students to continue learning languages in self access laboratories or centres. In contrast to classroom materials corpus-based learning materials in self access centres have be suited for self study without teacher instruction. But once students are familiar with the importance of frequency data and the use of concordancers they should be able to do exercises like the above-mentioned self-correction or hypothesis testing, and be able to work on semantic and thematic fields as well as search for domain-specific information which will be mentioned later. A process which is difficult to trigger and impossible to control is to get students to formulate their own research questions. This is only possible with very advanced students.

One tool which can be introduced in class and taken advantage outside of it is the use of internet corpora like the Business Letter Corpus or the Michigan Corpus of Academic Spoken English (MICASE). These can later serve as models of correct language use.

And finally, students should be made aware of the potential of the web as a corpus itself, of its advantages as well as of its drawbacks. As our students are much too often a bit too careless with information they find on the net, creating the right awareness of and skills for how to use the net is a delicate matter and demands for a good choice of examples and activities how to employ the web for language learning or hypothesis-testing purposes. As boot-strapping (the use of a specific computer program which uses seeds – special search terms – to build a corpus by the help of automated Google queries, cf. Baroni) might be beyond the scope of the normal language learner, the use of search engines can serve the same ends. Students can for instance search for certain multiword units if they want to verify a hunch. But they need to understand that search string results like "collect an experience" as a collocation is not acceptable and that they should be alarmed when the domains of the returned hits are predominantly German, Dutch and Swedish.

All in all, concordance techniques and related search techniques as well as models for verification of hunches are useful transferable skills which help students to become "media-literate" (cf. Rüschoff/Wolff 1999).

6.3.4 Domain-Specific Knowledge

Having dealt so far with mainly linguistic information retrieved by the help of corpus assistance I would now like to briefly deal with the possibility of investigating pieces of domain-specific knowledge from the corpus.

What Meyer and Macintosh say about terminographers and their use of corpora can be applied to teachers of a domain-specific language as well: "the terminographer uses the corpus not only as a source of linguistic evidence but also as a source of domain knowledge" (1996:278). Certain linguistic frames such as *is the, is a, such as*, etc. point to what Meyer and Macintosh call knowledge-rich contexts. For an overview of such lexical patterns introducing knowledge-rich contexts see Bowker and Pearson (2002:219). ENGICOR provides the following examples of domain-specific information:

Fig. 13: sample sentences containing knowledge-rich contexts

(1) A narrower [[**definition**]] of phytomining is the use of plants to obtain an economic return from metals extracted by a plant, whether from contaminated soils or from soils having naturally high concentrations of metals...
(2) In this sense, the term "reuse" is considered similar to "recycling", as [[**defined**]] in The American HeritageDictionary: "to extract and reuse (useful substances found in waste); to use again, esp. to reprocess in order to use again; to recondition and adapt to a new use or function."
(3) Appendix B - [[**Glossary**]] of Terms: Analyte - The chemical for which a sample is analyzed.
(4) A tub grinder [[**is a type of**]] hammer mill that uses a rotating tub intake system to crush wood and brush.
(5) Methanogens are [[**used for**]] the treatment of sludge in sewage treatment plants
(6) Surfactants are [[**used to**]] cause air bubbles to attach themselves to the valuable minerals, but not to the gangue.
(7) Phytoextraction has also been [[**referred to**]] as phytomining or biomining.
(8) Solvent extraction ([[**also known as**]] chemical extraction) is a cleanup method that uses solvents to extract or remove harmful chemicals from polluted materials.
(9) A trommel screen [[**consists of**]] a gently sloping, cylindrical screen that rotates.
(10) Free bromine [[**results from**]] the oxidation of the bromide ion in source water.

Search terms with the stem in *defin** or the word *glossary* provide pretty obvious hints for domain-specific knowledge (as shown in (1), (2), and (3)). Similar defining functions are also expressed by the search terms in (7) and (8). Other search terms (i.e. (4), (9), and (10)) indicate different kinds of relationships. Functions are described by the word material in (5) and (6).

What is exemplified by the sample sentences in Figure 15 is but the tip of the iceberg of domain-specific information found in a corpus. It is the kind of information which can be retrieved systematically. The major part cannot be retrieved systematically but only found in a concordance around a search term. Without knowing or understanding its specific technical implications, what I understand from the usage profile of *biodegradation* is that it can either occur *under aerobic* or *anaerobic conditions*. That there is to a certain degree *natural biodegradation* and if this is not sufficient it can be *enhanced* or *stimulated* and consequently *biodegradation rates* come into play. I can also learn from the usage profile that *biodegradation* is a process which happens predominantly on *landfills* to the organic fraction of waste and that there are also specific chemical substances which can be degraded like *benzene, alcohol* or *PCBs*. This information means that there is already a certain amount of conceptual information which might be vital in discussion processes in class if the teacher has to have at least a general idea about domain-specific concepts.

Tab. 34: usage profile of *biodegradation*

biodegradation (164x in corpus) noun rank 471 in keyword list	
N-1	**N+1**
aerobic biodegradation (11)	biodegradation process(es) (13)
anaerobic biodegradation (6)	biodegradation rate(s) (7)
natural biodegradation (10)	
enhance biodegradation (7)	
stimulate biodegradation (5)	
clusters	
biodegradation of (34) (alcohol/benzene/contaminants/landfills/organic waste/PCBs/these chemicals/...)	

6.3.5 Morphological Information

The various kinds of word-formation patterns active in environmental engineering English have been looked at in detail in Chapter 4. Here it was already stated that certain word-formation processes should be reflected in the ESP-course in order to enable students to first facilitate understanding of newly encountered words which they have not seen yet and secondly to keep up with language change in the further development of that variety.

Advanced students at university level already dispose of a considerable amount of implicit knowledge of word-formation patterns in the field of General English. By focussing on specific means of word formation which are particularly active in the domain-specific language, this implicit knowledge can be turned into explicit knowledge and thus be more readily applicable in language learning processes. This can be achieved by either teacher instruction or students' own research as was outlined under 6.3.2.2 above.

I claim that it is important for the students to understand that a specialist language is a dynamic process in which new words and word combinations come into use all the time designating new concepts of objects and processes originating from the progress in their field of study. Another important concept to understand for the students is that there is always the tendency to language economy while at the same time being as precise as possible. This can be exemplified by looking for instance at noun+noun compounds or complex adjectives and let students transform them into the longer syntactic phrases from which they might originate (i.e. *light-sensitive – sensitive to light*).

Chapter 4 has brought to light that not all word-formation processes are active to the same extent. Consequently there are areas on which classroom work should concentrate, whereas there are others which can be neglected. Above all, teaching and learning should focus on compounding and derivational processes as they are the ones which generate most new language items in environmental engineering English, followed by combining processes and conversion. As corpus work lends itself best to form-based searches, students can become active participants in searches based on word-formation patterns, coming to their own conclusions based on data retrieved. Objects of students research could for instance be series-forming elements typical for environmental engineering English like *–sensitive, -containing, -related*, the frequency and meaning potential of certain combining forms or affixes. To this end, I tried to establish a method of how to determine the productive derivational and combining elements. Some concepts like *bio-* and *phyto-* were found to have lent themselves to a considerable number of new items and are – as they are closely related to current technological trends in the domain - likely to continue being productive.

In the *–ables* case study (see 5.2.9.1.1.) it was shown in detail how new language items come into being by a process called conversion. The methodology applied in this paragraph is suitable for application in class as well which can further the students' understanding of how language underlies constant change. Being given items like *recyclables* or *biodegradables* students can undertake corpus searches in order to find the origins of such shortened concepts. Another idea which played a role in this paragraph was the fact that certain endings denote particular classes of objects. When either conversion or derivation generates have recently generated new members of these classes, then students should get familiar with their structure and meaning (i.e. *–ics* to denote chemical substances like *aliphatics, phenolics*, etc. or *–ants/-ents* to denote chemical agents like *adsorbants, coolants, flocculants*, etc.)

Less frequent word-formation processes like initialisms, clipping, blends, backformation, and word manufacture do not need to be subject to reflection in the ESP-course. Initialisms, clipped forms and blends are however very suitable objects for search tasks in the corpus and on the web and could be applied with interested students. Explaining the process of backformation though would be too complicated and the learning outcome does not justify the effort spent.

Examples of semantic change, as stated in 5.2.11, cannot systematically be retrieved with the help of an unannotated corpus, so only the findings encountered by chance can be used to illustrate language change evoked by that process. Still, students who are confronted by the task to speculate on why the word *pregnant* occurs quite often in ENGICOR and then search a concordance for the collocates of *pregnant* in environmental engineering English will most definitely understand how new meanings of words come into being.

Summing up, I would like to underline that instruction and own research of morphological characteristics of a domain-specific language can make students realise the information potential of lexical items and in this way acquire crucial skills for guessing strategies of unknown vocabulary and being able to cope with language change throughout their professional careers.

6.4 Summing Up

In this chapter I was trying to answer the questions as to whether the data extracted from a specialist corpus can serve to take "principled decisions" during all processes involved in teaching, starting with syllabus design and ending with the acquisition of strategies for life-long learning. In sum, having subjected each part of the teaching process to a careful examination on what additional advantage corpus data can provide, it can strongly be argued in favour of using corpus data. Its biggest advantage lies in the objective method of corpus analysis the linguistic data derives from. Empirical evidence proves what the most salient language items are and further analysis at subcorpus and text level provide more insights into how to structure the data.

Moreover, a corpus cannot only help in structuring the amount of data but also allows for immediate application in learning processes, be it consciously followed up by students or implicitly intertwined in all kinds of classroom processes. Bringing corpus data into the classroom creates a learning atmosphere where students are actively involved in investigation processes. They are guided to reflecting on specialist language, a result which meets the demand of university level ESP-courses and which goes well beyond a purely communicative approach.

7 Recap

To conclude this study, all that remains to do is to see which of the above-established aims could be achieved and to summarize which findings and insights were deemed most productive for the purposes of the design of corpus-informed language courses.

At the origin of this piece of research was the perception of a lack of suitable teaching and learning material for the specialist language course of environmental engineering English at university level as well as the need to operationalise preparation for the course in terms of syllabus, materials and in-class activities.

As a means of deriving the necessary information, a small specialist corpus of 2 million running words from engineering texts was created with the aim to test whether it could provide the lacking information. This was done based on the firm conviction that it is the phrasal character of language which should determine the way a language is taught. Consequently, the corpus was searched for information on extended units of meaning, i.e. collocational and colligational behaviour as well as information on semantic preference and semantic prosody, in order to create a basis for students to enable them to get as close as possible to native-like fluency in the domain-specific language.

The foremost aim was to adapt the different levels at which extended units of meaning manifest themselves to the use of a domain-specific language in order to describe environmental engineering English with a view to pedagogical applications. As a practical sub-aim, I wanted to create a bank of linguistic information about selected items of this language. Furthermore, as a methodological sub-aim, I wanted to develop an operational procedure for deducing the sought information from this inventory of linguistic information. All of this was done by consistently checking against established corpus linguistics knowledge in order to use ENGICOR as a testbed for the usefulness and applicability of basic corpus linguistics principles in a domain-specific language. All in all, the present study is a small corpus investigation with a view to applying language data retrieved for teaching and learning processes. In the following paragraphs I will explain in summarized form how the aims were fulfilled.

Methodology-wise, the first step was to create the corpus, ENGICOR, based on information from subject-field specialists. Then a keyword analysis was done to arrive at a subset of words which were subjected to further analysis by concordancing and thus deducting information on collocation, colligation, semantic preference and semantic prosody which was then documented and gathered in the usage profiles, a kind of inventory for keeping the corpus information. Afterwards, the different subcorpora were searched for typical words and the importance of keyword analyses at text levels were considered.

It needs to be mentioned that a small-scale study corpus like ENGICOR is not an ideal object of research. As a one-person research endeavour, I was only able to include written texts from the internet and from a textbook. It had furthermore to be a static corpus rather than a monitor corpus, a drawback that was felt when investigating certain word-formation patterns. In order to balance out the disadvantages, special care went into choosing texts from a wide range of topics and text types and authenticity and as well as the current relevance of texts were considered to be of paramount importance. With these shortcomings in mind, investigations were undertaken into the question to what degree the findings from this study can be generalised to the whole domain. For that purpose, a mini-corpus as a kind of comparable corpus was created and in the analysis, it was found that at word level, the findings are indeed representative, whereas at multiword-level the statements can only be confidently made for environmental engineering English in the context of the United States, as ENGICOR mainly consists of American English texts.

Moreover, with an analysis of how much text is covered by all words contained in the General Service List, the Academic Word List and of the first 500 keywords from ENGICOR, it was found that it is well worth concentrating on them as they together cover 94.3% of the texts in the corpus. This proved first that environmental engineering English is – compared to General English – a very restricted language, that is to say a language with a limited amount of lexical items and also shows that concentrating in class on the first 500 keywords can be justified in terms of the effort spent.

In order to have the corpus information at hand for the linguistic analysis as well as for its use in class, the usage profiles were designed to store corpus data. An analysis of the collocational behaviour of the subgroup of chemical nouns (i.e. *aluminium, carbon, lead, nitrogen, phosphorus*, etc.), which revealed a lot of shared collocational frames, led to the question whether a similar behaviour could also be found for other subgroups. It was hoped that on top of information on semantic preference, which shows a kind of orderly behaviour of some semantic groups, a similar behaviour could be detected for topic-based subgroups as for instance the group of recyclable materials (i.e. *glass, rubber, plastics, wood, paper, metal*, etc.). It was speculated that in such a case, rather than teaching the collocational profile of *glass*, and that of *plastic* and that of *paper*, etc., it would be possible to establish a superordinate category of the collocational profile of recyclable materials. It was however found that this hope could not be fulfilled if we stick to prioritising the principle of importance of frequency information. The collocational profiles of the different members of the class "recyclable materials" were different from each other to a point that no sensible conclusions could be drawn in terms of a superordinate collocational pattern.

Specifying the work that was begun in the detailed consistency analysis of the word materials of the different subcorpora, an analysis into the key keywords of the subcorpora and their associates was undertaken. It was found that they are a valuable first indication for topic-related language material on whose collocational environment it should first be concentrated when embarking on a specific topic like mining reclamation. This should be done before starting to work on the particular keywords of the single text. Key keywords were found to create a network of intercollocation; that is to say words which we can expect with high probability to occur in a text of a certain topic.

Particular emphasis was put on word-formation patterns typical for environmental engineering English. Language change is induced by the need to designate new concepts in emerging technologies. To this end, recombinations and derivations from the existing word material usually happen rather than simply creating new words.

The investigation into the typical word-formation patterns was done with the following aims in mind: students should be furnished with a sound basis of morphological knowledge in order to on the one hand be able to use this knowledge in processes of understanding unknown vocabulary and thus making use of the meaning load word-formation processes can offer, and on the other hand to prepare the ground for making sense of future language change, as environmental engineering as an innovative branch is expected to develop further and thus creating even more need to designate new concepts. Here again the information collected in the usage profiles came into play and generalisations were made on the basis of this documented corpus information. All word-formation phenomena were investigated in terms of which elements combine or are involved in the process and what the results of these processes are.

To this effect, compounding patterns were considered first. A categorisation of the found patterns was established and their semantics considered. It was for instance found that most noun-centred compounds could be assigned to one of the following semantic categories: *processes and operations, process indicators, process agents and products, process equipment and devices* and *process locations*. It was found that this very much reflected the world of environmental engineering and helps in recontextualisation activities to concentrate on the situational framework within which environmental engineering is operating. Of special interest in the field of compounding processes were series-forming elements restricted by very obvious semantic preferences and semantic prosodies. As they designate precise, shortened concepts used for premodifying purposes (i.e. *area-intensive construction, metal-bearing solution, VOC-laden fumes*, etc.) it is worth concentrating on them in the ESP-class.

The inventory of compounding patterns was created to show which phenomena are most frequent and how they stand for shortened concepts. Noun compounds were most frequent.

With combining processes it was found that is necessary for students to understand the meaning load added by the different combining forms and to concentrate on those which are particularly active in producing new words. This always has to be considered together with the development of the emerging technologies upon which their existence lies, as they can be assumed to bring about more innovations in the future. When looking at the productivity of single word-formation processes, quantitative and qualitative methods were combined in order to determine which combining forms might yield further new words in the future. Information of purely quantitative methods like productivity in the narrow sense were not found reliable enough for such a small corpus. It was moreover found that a combination of OED data together with Google searches could reliably show whether a certain item can be considered a relatively new phenomenon of which to expect further developments.

In the paragraph about derivational patterns in environmental engineering English, the focus was again to create an inventory of the means by which new words are produced adding a prefix or a suffix to already existing words. Here as well investigations into the productivity of selected noun, adjective and verb endings were undertaken.

Minor word-formation processes like initialisms, clipping, backformation, blends, word manufacture and loan words as well as processes of semantic change were exemplified by typical corpus findings.

Conversion as one further process of producing new words was then focused on in detail as it was found that at least conversion processes where adjective-noun compounds change into nouns based on the involved adjectives are a field which did not get the attention it deserves in the special literature. To this end, the *–ables* case study revealed in detail how collective terms of the form adjective+noun are shortened to a noun based on the adjective as in *biodegradable materials > biodegradables*. It was furthermore aimed at demonstrating that this phenomenon is not restricted to the ending *–ables* but concerns other endings as well to a similar degree and for similar reasons. A view of the semantic features that such classes of new words have in common proved to be vital information for learning purposes as students should be able to understand that in case they encounter a new word in *–ant(s)*, it is very likely that this word then designates a chemical agent (i.e. *leaching agent(s) – leachant(s))*. It was found that the discussed conversion processes are part of an active restructuring of syntax, a fact which supports the development to shorten concepts if the context supports the clarity of the meaning information given.

It needs to be emphasized at this point, that when it is said that students should be familiarised with word-formation patterns which predominantly produce new words in the domain, it is not meant that students should become linguistic experts in this field. I am nevertheless convinced that by fostering reflection on current word-formation processes, students can be sensitized towards the dynamic character of language.

The point where both strands of investigation in this study – the importance of collocational and that of morphological information, i.e. the information from the environment surrounding a word and the word structure itself - are brought together is that both are expected to help the students in the understanding of language and in their own production of it. The next question that had to be answered then was how the information could be made use of for preparing the teaching and in class.

For that purpose a method was developed of how to structure the linguistic data found from corpus investigation. This was at the most global level the 500 keywords and at the most detailed the single usage profiles. In between these two levels, it was of importance which words relate to certain subtopics, and which occur evenly throughout the whole corpus as well as the relevance of keywords of single texts. This information could be established as useful for the pre-classroom use – that is in syllabus and materials design – in order to help making informed decisions at this stage. What needs to be stressed here is the concept of recontextualisation. Corpus data was condensed into the usage profiles but needs to be extracted again for the use in teaching and learning processes. The way to put this process of recontextualisation into action is a crucial one as it is at this stage where linguistic data needs to be transferred into classroom activities. Suitable classroom activities need to be carefully planned in order to ensure that students can grasp the information of an extended unit of meaning in its entirety. For in-class activities, two routes to put this into action were considered: firstly the route of priming - that is creating repeated encounters with relevant language material on a more unconscious level – and secondly educational linguistics, under which I subsume all teacher-initiated activities which lead to reflection on language data as well as enriching students' own research with the help of the corpus.
Unlike approaches which focus exclusively on priming, the inclusion of educational linguistics in the teaching and the learning of a specialised language furthers independent learning beyond the classroom. To this end, students should also be familiarized with the potential of freely available internet corpora and with the scope and dangers of the web as a kind of corpus.

It can definitely be claimed at the end of this study that corpus work enables teachers to embark on their own research in applied linguistics, providing such a vast array of very much needed information in the process of ESP-teaching. This information serves to focus the view on the collocational nature of language, a prerequisite for transferring this information into the language learning process.

In conclusion, this study aimed at describing the specialist language of environmental engineering English in order to discuss the relevance of the established language data for classroom use. It can conclusively be emphasised that corpus work and the chosen methodology made it possible to realize these targets as well as made substantial contributions to the description and organisation of the course syllabus as well as to the specific materials and activities which could be devised. Suitable operational procedures for the study of words in context were established and should in this way enable students to use language items in correct settings to make ESP-courses of real benefit to them.

What is now left to be envisaged after this study is putting the established language data and framework of methodological steps into action when preparing the course and during the teaching and learning processes. The claims of the present approach could as well be tested in terms of the learning outcome of the students compared to other approaches.

It would furthermore be of interest to see how the findings from this study translate to a domain-specific language other than environmental engineering English, be it another technical language or another variety.

References

Adams, V. (2001). *Complex Words in English.* Harlow: Pearson Education.

Algeo, J. (1980). Where do all the new words come from. *American Speech,* 55, 264-277.

Aston, G. (1997). Corpora in Language Pedagogy: Matching Theory and Practice. In G. Cook and B. Seidlhofer (Eds.), *Principle and Practice in Applied Linguistics.* Oxford: Oxford University Press, 257-270.

Atkins, S., Clear, J. & Ostler, N. (1992). Corpus Design Criteria. *Literary and Linguistic Computing* , 7(1), 1-16.

Baayen, R. H. & Renouf, A. (1996). Chronicling The Times: Productive Lexical Innovations in an English Newspaper. *Language,* 72, 69-96.

Baayen, R.H. (1991). Quantitative Aspects of Morphological Productivity. In G.E. Booij and J. van Marle (Eds.), *Yearbook of Morphology 1991.* Dordrecht: Kluwer Academic Publishers, 109-149.

Baayen, R.H. (1992). On Frequency, Transparency, and Productivity. In G.E. Booij and J. van Marle (Eds.), *Yearbook of Morphology 1992.* Dordrecht: Kluwer Academic Publishers, 181-208.

Bahns, J. (1993): Lexical Collocations. *ELT Journal,* 47(1), 56-63.

Bahns, J. & Eldaw, M. (1993). Should We Teach EFL Students Collocations? *System,* 21(1), 101-114.

Barnhart, C.L. et al. (1973). *The Barnhart Dictionary of New English since 1963.* Bronxville (N.Y).: Barnhart/Harper and Row.

Baroni, M. & Bernardini, S. (2004). *BootCat: Bootstrapping Corpora and Terms from the Web.* Retrieved from http://sslmit.unibo.it/~baroni/publications/lrec2004/bootcat_lrec_2004.pdf, as of 09 February 2006.

Bauer, L. & Huddleston, R. (2002). Lexical Word-Formation. In R. Huddleston and G.K. Pullum (Eds.), *The Cambridge Grammar of the English Language.* Cambridge: Cambridge University Press, 1623-1719.

Bauer, L. & Renouf, A. (2001). A Corpus-Based Study of Compounding in English. *Journal of English Linguistics,* 29(2), 101-123.

Bauer, L. (1983). *English Word-Formation.* Cambridge: Cambridge University Press.

Bauer, L. (1992). Scalar Productivity and –lily Adverbs. In G.E. Booij and J. van Marle (Eds.), *Yearbook of Morphology 1991.* Dordrecht: Kluwer Academic Publishers, 185-191.

Bauer, L. (1997). *Watching English Change.* London: Longman.

Bauer, L. (1998). When is a Sequence of Two Nouns a Compound in English. *English Language Lingustics*, 2(1), 65-86.

Benson. M. et al. (1997). *The BBI Dictionary of English Word Combinations*. Amsterdam: Benjamins.

Bernardini, S. (2000). Systematising Serendipity – Proposals for Concordancing Large Corpora with Language Learners. In L. Burnard and T. McEnery (Eds.), *Rethinking Language Pedagogy from a Corpus Perspective – Papers from the Third International Conference on Teaching and Language Corpora*. Frankfurt/a.M.: Peter Lang, 183-190.

Biber, D. et al. (1999). *Longman Grammar of Spoken and Written English*. Harlow: Pearson Education.

Bondi, M. (2001). Small Corpora and Language Variation: Reflexivity Across Genres. In M. Ghadessy et al. (Eds.), *Small Corpus Studies and ELT: Theory and Practice*. Amsterdam: Benjamins, 135-174.

Bowker, L. & Pearson, J., (2002). *Working with Specialized Language – A Practical Guide to Using Corpora*. London and New York: Routledge.

Bowker, L. (2000). Towards a Methodology for Exploiting Specialized Target Language Corpora as Translation Resources. *International Journal of Corpus Linguistics*, 5(1), 17-52.

Brindley, G. (1989). The Role of Needs Analysis in Adult ESL Design. In K. Johnson (Ed.), *The Second Language Curriculum*. Cambridge: Cambridge University Press.

Cabré, M.T. (1998). *Terminology – Theory, Methods and Applications*. Amsterdam: Benjamins.

Caso, A.L. (1980). The Production of New Scientific Terms. *American Speech*, 55, 101-111.

Clear, J. (1994). I can't see the sense in a large corpus. *Papers in Computational Lexicography. Proceedings of Complex 99*, Budapest, 33-48.

Collocott, T.C. (Ed.) (1961). *Dictionary of Science and Technology*. Edinburgh: W. & R. Chambers.

Coxhead, A. (2000). A New Academic Word List. *TESOL Quarterly*, 34(2), 213-239.

Curado, A. (2001). Lexical Behaviour in Academic and Technical Corpora: Implications for ESP Development. *Language Learning and Technology*, 5(3), 106-129.

Curado, A. (2002). *A Lexical Common Core in English for Information Science and Technology*. Cáceres: Servicio de Publicaciones de la Universidad de Extremadura.

Danielsson, P. & Mahlberg, M. (2003). There is more to knowing a language than knowing its words: Using parallel corpora in the bilingual classroom. *ESP-World*, 3(6-2), retrieved at www.esp-world.info/articles_6/DanielssonMahlberg2003.htm, as of 10 February 2006.

Detering, K. (2000). Wortverschmelzungen – Semantische Untersuchungen und didaktische Überlegungen. In K. Detering (Ed.), *Wortschatz und Wortschatzvermittlung: Linguistische und Didaktische Aspekte*, Frankfurt/a.M.: Peter Lang, 89-125.

Erman, B. & Warren, B. (2000). The Idiom Principle and the Open Choice Principle. *Text* 20(1), 29-62.

Fischer, R. (1998). *Lexical Change in Present-Day English.* Tübingen: Gunter Narr Verlag.

Flowerdew, J. (2001). Concordancing as a Tool in Course Design. In M. Ghadessy et al. (Eds.), *Small Corpus Studies and ELT: Theory and Practice.* Amsterdam: Benjamins, 71-92.

Flowerdew, L. (2001). The Exploitation of Small Learner Corpora in EAP Materials Design. In M. Ghadessy et al. (Eds.), *Small Corpus Studies and ELT: Theory and Practice.* Amsterdam: Benjamins, 363-380.

Fluck, H.R. (1997). *Fachdeutsch in Naturwissenschaft und Technik: Deutsch als Fremdsprache.* Heidelberg: Julius Groos Verlag.

Ghadessy, M. et al. (2001). *Small Corpus Studies and ELT: Theory and Practice.* Amsterdam: Benjamins.

Ghadessy. M. and Gao, Y. (2001). Small Corpora and Translation: Comparing Thematic Organization in Two Languages. In M. Ghadessy et al. (Eds.), *Small Corpus Studies and ELT: Theory and Practice.* Amsterdam: Benjamins, 335-362.

Gledhill, C. (2000). *Collocations in Science Writing.* Tübingen: Gunter Narr Verlag.

Halliday, M.A.K. & Hasan, R. (1976). *Cohesion in English.* London: Longman.

Hänchen, R. (2002). *Die Französische Marketingsprache – Eine diachrone Untersuchung ihrer Terminologie anhand der Revue Française du Marketing (1960-2000).* Frankfurt/a.M.: Peter Lang.

Hansen, B. et al. (1990). *Englische Lexikologie: Einführung in Wortbildung und lexikalische Semantik.* Leipzig: VEB Verlag Enzyklopädie.

Henry. A. & Roseberry, R.L. (2001). Using a Small Corpus to Obtain Data for Teaching a Genre. In M. Ghadessy et al. (Eds.), *Small Corpus Studies and ELT: Theory and Practice.* Amsterdam: Benjamins, 93-134.

Hickey, R. (2003). Tracking Lexical Change in Present-Day English.In A. Wilson et al. (Eds.), *Corpus Linguistics by the Lune: A Festschrift for Geoffrey Leech.* Frankfurt/a.M.: Peter Lang, 93-105.

Higgins, J. & Johns, T. (1984). Computers in Language Learning, London: Collins ELT, 88-93.

Hill, J. & Lewis, M. (1999). *Dictionary of Selected Collocations.* Hove: LTP.

Hoey, M. (2005). *Lexical Priming – A New Theory of Words and Language.* London and New York: Routledge.

Huddleston, R. & Pullum, G.K. (2002). *The Cambridge Grammar of the English Language.* Cambridge: Cambridge University Press.

Hunston, S. & Thompson, G. (Eds.) (2000). *Evaluation in Text: Authorial Stance and the Construction of Discourse.* Oxford: Oxford University Press.

Hunston, S. et al. (1997). Grammar and Vocabulary: Showing the Connections. *ELT Journal,* 51(3), 208 – 217.

Johns, T. (1986). Microconcord – A Language Learner's Research Tool. *System,* 14(2), 151-162.

Johns, T. (1988). Whence and Whither Classroom Concordancing? In T.P. Bongaerts et al. (Eds.), Computer Applications in Language Learning (Foris), 9-27.

Kettemann, B. (n.d.). *On the Role of Context in Syntax and Semantics.* Retrieved at http://www-gewi.kfunigraz.ac.at/project/corpus.html., on 09 February 2006.

Laufer, B. (1988). What Percentage of Text Lexis is Essential for Comprehension?" In C. Lauren, & M. Nordman (Eds.), *Special Language: From Humans Thinking to Thinking Machines.* Clevedon: Multilingual Matters, 316-323.

Lawson, A. (2001). Collecting, Aligning and Analysing Parallel Corpora. In M. Ghadessy et al. (Eds.), *Small Corpus Studies and ELT: Theory and Practice.* Amsterdam: Benjamins, 279-310.

Lawson, A. (1997). English Words in Use – Compiling a Dictionary of Collocation. In B. Lewandowska-Tomaszczyk (Ed.) *International Conference on Practical Applications in Language – PALC 1997.* Frankfurt/a.M.: Peter Lang, 225-235.

Lehrer, A. (1996). Identifying and Interpreting Blends: An Experiment Approach. *Cognitive Linguistics,* 4-7, 359-390.

Lewis, M. (1993). *The Lexical Approach – The State of Art of ELT and the Way Forward.* Hove: LTP.

Lewis, M. (1997). *Implementing the Lexical Approach – Putting Theory into Practice.* Hove: LTP.

Lewis, M. (2000). *Teaching Collocation – Further Developments in the Lexical Approach.* Hove: LTP.

Liu, J. & Nesi, H. (1999). Are We Teaching the Right Words? A Study of Students' Receptive Knowledge of Two Types of Vocabulary: 'Subtechnical' and 'Technical'. In H. Bool, H. & P. Luford. (Eds.). *Academic Standards and Expectations: the role of EAP.* Nottingham: Nottingham University Press.

Ma, B.K.C. (1993). Small-Corpora Concordancing in ESL Teaching and Learning. *Hong Kong Papers in Linguistics and Language Teaching,* 16, 11-30.

Marchand, H. (1969). *The Categories and Types of Present-Day English Word-Formation.* München: C.H. Beck´sche Verlagsbuchhandlung.

Meyer, I. & MacIntosh, K. (1996). The Corpus from a Terminographer´s Viewpoint. *International Journal of Corpus Linguistics,* 1(2), 257-285.

Mukherjee, J. (2002). *Korpuslinguistik und Englischunterricht: Eine Einführung*. Frankfurt/a.M.: Peter Lang.

Nation, P. (2001). Using Small Corpora to Investigate Learner Needs: Two Vocabulary Research Tools. In M. Ghadessy et al. (Eds.), *Small Corpus Studies and ELT: Theory and Practice*. Amsterdam: Benjamins, 31-46.

Nelson, M. (2000). *A Corpus-Based Study of Business English and Business English Teaching Materials.* PhD thesis, University of Manchester. Retrieved from http://www.kielikanava.com/thesis.html, as of 10 March 2005.

Nuccorini, S. (2002). *Phrases and Phraseology – Data and Descriptions*. Bern: Peter Lang.

Palmer, F.R. (1968)*: Selected Papers of J.R.Firth 1952-59*. London: Longman.

Partington, A. (1998) *Patterns and Meanings*. Amsterdam:Benjamins.

Partington. A. (2004). Utterly Content in Each Other's Company - Semantic Prosody and Semantic Preference. *International Journal of Corpus Linguistics*, 9(1), 131-156.

Pawley, A. & Syder, F.H. (1983): Two Puzzles for Linguistic Theory: Nativelike Selection and Nativelike Fluency. In J. Richards and R. Schmidt (Eds.), *Language and Communication*. London: Longman, 191-227.

Pérez-Paredes, P. (2003). *Small Corpora as Assisting Tools in the Teaching of English News Language: A Preliminary Tokens-Based Examination of Michael Swan's Practical English Usage News Language Wordlist*. Retrieved from www.esp-world.info/articles_6/pascual.htm, as of 10 October 2005.

Phillips, M. (1985). *Aspects of Text Structure – An Investigation of Lexical Organisation of Text*. Amsterdam: Elsevier.

Plag, I. (1999). *Morphological Productivity – Structural Constraints in English Derivation*. Berlin: Mouton de Gruyter.

Plag, I. (2003). *Word-Formation in English*. Cambridge: Cambridge University Press.

Quirk, R. et al. (1985). *A Comprehensive Grammar of the English Language*. London: Longman.

Reilly, M. (2002). *Toronto Sludge Fertilizer Ordered Removed from Farm*. Retrieved from http://list.web.net/archives/sludgewatch-l/2002-February.txt, as of 09 September 2005.

Renouf, A. (1993). A Word in Time: First Findings from the Investigation of Dynamic Text. In J. Aarts et al. (Eds.), *English Language Corpora: Design, Analysis and Exploitation*, Amsterdam: Rodopi, 279-288.

Rundell, M. (2000). TheseThings Happen. *Humanising Language Teaching,* 2(4). Retrieved from www.hltmag.co.uk/jul00/idea.htm, as of 24 September 2005.

Rüschoff, B. & Wolff, D. (1999). *Fremdsprachenlernen in der Wissensgesellschaft – Zum Einsatz der Neuen Technologien in Schule und Unterricht.* Ismaning: Max Hueber Verlag.

Sager, J.C. (1997). Term Formation. In S.E. Wright and G. Budin (Eds.), *Handbook of Terminology Management,* , Amsterdam: Benjamins, Vol. 1, 25-41.

Scott, M. (2001). Mapping Key Words to Problem and Solution. In M. Scott and G. Thompson (Eds.); *Patterns of Text: In Honour of Michael Hoey.* Amsterdam: Benjamins, 109-127.

Siepmann, D. (2004). Kollokationen und Fremdsprachenlernen: Imitation und Kreation, Figur und Hintergrund. *Praxis Fremdsprachenunterricht,* 2(2004), 107-113.

Sinclair, J. (1982). Reflections on Computer Corpora in English Language Research. In S. Johansson (Ed.), *Computer Corpora in English Language Research.* Bergen: NAVF.

Sinclair, J. (1987). Collocation: A Progress Report. In R. Steele. & T. Threadgold (Eds.), *Language Topics.* Amsterdam: Benjamins, 319-331.

Sinclair, J. (1995). *Corpus, Concordance, Collocation.* Oxford: Oxford University Press.

Sinclair, J. (1996a). The Search for Units of Meaning. *Textus.* IX, 75-106.

Sinclair, J. (1996b). The Empty Lexicon. *International Journal of Corpus Linguistics,* 1(1), 99-119.

Sinclair, J. (1999). Tuscan Word Centre Website, retrieved at www.twc.it/happen.html, as of 12 May 2007.

Sinclair, J. (2001). Preface to Ghadessy et al. (Eds.), *Small Corpus Studies and ELT: Theory and Practice.* Amsterdam: Benjamins.

Sinclair, J. (2004). Corpus and Text – Basic Principles. In M. Wynne (Ed.), *Developing Linguistic Corpora: A Guide to Good Practice.* Retrieved from www.ahds.ac.uk/creating/ guides/linguistic-corpora/chapter1.htm, as of 6 January 2006.

Sinclair, J. and A. Renouf (1988). A Lexical Syllabus for Language Learning. In Carter R. and M. McCarthy (Eds.), *Vocabulary and Language Teaching.* Harlow: Longman, 140-158.

Smadja, F. (1993). Retrieving Collocations from Text: Xtract. *Computational Linguistics,* 19(1), 143-177.

Stubbs, M. (1997). Cause and Provide. In G. Cook and B. Seidlhofer (Eds.), *Principle & Practice in Applied Linguistics.* Oxford: Oxford University Press, 247-256.

Stubbs, M. (2001). *Words and Phrases : Corpus Studies of Lexical Semantics.*Oxford: Blackwell.

Stubbs, M. (2002). Two Quantitative Methods of Studying Phraseology in English. *International Journal of Corpus Linguistics,* 7(2), 215-244.

Stubbs, M. and Barth, I. (2003). Using Recurrent Phrases as Text-Type Discriminators. A Quantitative Method and Some Findings. *Functions of Language*. 10(1), 61-104.

Teubert, W. (1998). *Neologie und Korpus*. Tübingen: Gunter Narr Verlag.

Teubert, W. (1999). Korpuslinguistik und Lexikographie. *Deutsche Sprache: Zeitschrift für Theorie, Praxis, Dokumentation,* 27(4), 292-313.

Thompson, G. (2001). Corpus, Comparison, Culture: Doing the Same Things Differently in Different Cultures. In M. Ghadessy et al. (Eds.), *Small Corpus Studies and ELT: Theory and Practice*. Amsterdam: Benjamins, 311-334.

Tognini-Bonelli, E. (2000). Corpus Classroom Currency. *Naujoji Metodologija. Darbai ir Dienos*. Retrieved from http://donelaitis.vdu.lt/publikacijos/bonelli.pdf., as of 06 January 2006.

Tognini-Bonelli, E. (2001). *Corpus linguistics at work*. Amsterdam: Benjamins.

Tognini-Bonelli, E. (2002a). Functionally Complete Units of Meaning Across English and Italian: Towards a Corpus-Driven Approach. In B. Altenberg & S. Granger (Eds.) *Lexis in Contrast – Corpus-Based Approaches*. Amsterdam: Benjamins.

Tognini-Bonelli, E. (2002b). Between Phraseology and Terminology in the Language of Economics. In S. Nuccorini (Ed.), *Phrases and Phraseology – Data and Descriptions*. Bern: Peter Lang, 65-84.

Tribble, C. (1997). Improvising Corpora for ELT: Quick and Dirty Ways of Developing Corpora for Language Teaching. Retrieved from http://www.ctribble.co.uk/text/Palc.htm, as of 11 October 2005.

Tribble, C. (2000). *Genres, Keywords, Teaching: Towards a Pedagogic Account of the Language of Project Proposals*. Retrieved from http://www.ctribble.co.uk/text/Genre.htm, as of 11 October 2005.

van Marle, J. (1991). The Relationship Between Morphological Productivity and Frequency: A Comment on Baayen's Performance-Oriented Conception of Morphological Productivity. In G.E. Booij and J. van Marle (Eds.), *Yearbook of Morphology 1991*. Dordrecht: Kluwer Academic Publishers, 151-163.

Weiner, R. & Matthews, R. (2003). *Environmental Engineering*. Newton (MA): Butterworth-Heinemann.

West, M. (1953). *A General Service List of English Words*. London: Longman.

Williams, G. (2002). In Search for Representativity in specialised Corpora – Categorisation through Collocation. *International Journal of Corpus Linguistics,* 7(1), 43-64.

Williams, G.C. (1999). *Les Réseaux Collocationnels dans la Construction et l'Exploitation d'un Corpus dans le Cadre d'une Communauté de Discours Scientifique*. PhD thesis. Retrieved from http://perso.wanadoo.fr/geoffrey.williams/thesis/, as of 10 March 2005.

Willis, D. (1990). *The Lexical Syllabus: a New Approach to Language Teaching*. London & Glasgow: Collins ELT.

Wilson, A. et al. (Eds.) (2003). *Corpus Linguistics by the Lune: A Festschrift for Geoffrey Leech*. Frankfurt/a.M.: Peter Lang.

Wray, A. & Perkins, M.R. (2000). The Functions of Formulaic Language: An Integrated Model. *Language & Communication,* 20, 1-28.

Software

Cobb, T. (n.d.). Vocab Profile Web VP (v 2.0). retrieved at www.lextutor.ca/vp/eng/, as of 6 January 2006.

MonoConcPro – information about the concordance program can be found at www.athel.com, retrieved 26 January 2006.

OED - Oxford English Dictionary (Second Edition) on CD-ROM version 3.0.

WordSmith 4 – information about the program can be found at www.lexically.net/wordsmith/version4/, as of 10 March 2005.

Online-Corpora/Access to Information on Corpora

British National Corpus – information on this corpus can be found at www.natcorp.ox.ac.uk/, as of 10 March 2005.

VIEW - British National Corpus – interface, by Mark Davies, retrieved at http://view.byu.edu/, as of 22 February 2006.

Business Letter Corpus, retrieved at http://ysomeya.hp.infoseek.co.jp, as of 09 February 2006.

Michigan Corpus of Academic Spoken English. Retrieved at http://micase.umdl.umich.edu/m/micase/, as of 09 February 2006.

Appendices

Appendix A – Lists

1. Keyword List – The 500 first keywords from the ENGICOR-BNC-comparison

1	WASTE	51	METALS	101	REACTIVE
2	WATER	52	OZONE	102	DISCHARGE
3	TAILINGS	53	MINE	103	RESULTS
4	WASTEWATER	54	MONITORING	104	CLEANUP
5	SITE	55	INJECTION	105	DISSOLVED
6	TREATMENT	56	FILTER	106	OPERATION
7	SOIL	57	CONTAMINATION	107	REDUCTION
8	LANDFILL	58	RUNOFF	108	SITES
9	GROUNDWATER	59	DISINFECTION	109	DATA
10	MATERIAL	60	TYPICALLY	110	FEEDSTOCK
11	CONCENTRATIONS	61	NITROGEN	111	MSW
12	CONTAMINANT	62	OXYGEN	112	SUBSURFACE
13	FLOW	63	SITU	113	LEACH
14	CONTAMINANTS	64	LANDFILLS	114	PUMP
15	EFFLUENT	65	POTENTIAL	115	MAINTENANCE
16	RECYCLING	66	NITRIFICATION	116	COLOR
17	CHLORINE	67	MINING	117	DETERMINE
18	SOLIDS	68	WELLS	118	PARTICLE
19	FACILITY	69	UV	119	TECHNOLOGIES
20	ENVIRONMENTAL	70	TEMPERATURE	120	GLASS
21	MATERIALS	71	CARBON	121	REQUIRED
22	ORGANIC	72	AEROBIC	122	STREAM
23	PROCESS	73	APPROXIMATELY	123	PROCESSES
24	CONCENTRATION	74	PRODUCT	124	ORGANICS
25	pH	75	LEACHING	125	CONTROL

26	RECYCLED	76	REUSE	126	PHYTOREMEDIATION
27	SURFACE	77	RATE	127	PRECIPITATION
28	CONTAMINATED	78	PRODUCTS	128	CONDITIONS
29	REMOVAL	79	AMMONIA	129	PLASTICS
30	OXIDATION	80	LEACHATE	130	COMBUSTION
31	PLANT	81	EMISSIONS	131	TEST
32	HYDRAULIC	82	HDPE	132	DRAINAGE
33	COMPOST	83	TIRES	133	FIBER
34	WASTES	84	TCE	134	PLUME
35	REMEDIATION	85	PLASTIC	135	SEWER
36	CHEMICAL	86	EXTRACTION	136	OPERATIONS
37	COMPOSTING	87	ADDITION	137	DIOXIDE
38	SAND	88	IMPOUNDMENT	138	VOLUME
39	HAZARDOUS	89	ODOR	139	CONTAINERS
40	COPPER	90	SOLID	140	SULFIDE
41	DISPOSAL	91	TIRE	141	FIGURE
42	REACTOR	92	PARTICLES	142	BIOREMEDIATION
43	AIR	93	RUBBER	143	CONSTRUCTION
44	ORE	94	ANAEROBIC	144	PARAMETERS
45	SLUDGE	95	BIOSOLIDS	145	REQUIREMENTS
46	SAMPLES	96	SAMPLING	146	CONDUCTED
47	AQUIFER	97	GAS	147	SEEPAGE
48	ACID	98	COD	148	SOLUTION
49	EQUIPMENT	99	RECOVERY	149	SOILS
50	COMPOUNDS	100	ZONE	150	POLLUTANTS
151	SBR	201	SEPTIC	251	DENSITY
152	QUALITY	202	VAPOR	252	EFFICIENCY
153	MUNICIPAL	203	CLAY	253	SCALE
154	METAL	204	URANIUM	254	MIXING
155	TECHNOLOGY	205	INSTALLED	255	TANK
156	CONTENT	206	TEMPERATURES	256	DISINFECTANT

157	PROCESSING	207	BATCH	257	CHARACTERISTICS
158	LOCATED	208	CHARACTERIZATION	258	GAC
159	DEGRADATION	209	POLLUTANT	259	SIZE
160	STORAGE	210	MASS	260	PERMANGANATE
161	METHODS	211	COLLOID	261	DOWNGRADIENT
162	SPECIFIC	212	PERMEABILITY	262	EFFECTIVE
163	COLLECTED	213	LIME	263	INCLUDE
164	INFLUENT	214	ADSORPTION	264	BOD
165	TESTING	215	POND	265	SEDIMENTS
166	REDUCE	216	FLOWS	266	OPERATING
167	LOADING	217	TRACER	267	FILTRATION
168	FLOTATION	218	MICROBIAL	268	OXIDIZED
169	ANOXIC	219	PER	269	BASIN
170	PHASE	220	GENERATION	270	POLLUTION
171	BIOMASS	221	MILL	271	FIBERS
172	MICROORGANISMS	222	LEVELS	272	BIOLOGICAL
173	DENITRIFICATION	223	FOUNDRY	273	FEED
174	PERFORMANCE	224	ACTIVATED	274	PUMPS
175	SEDIMENT	225	PUMPING	275	REMEDIAL
176	INACTIVATION	226	AQUEOUS	276	REMOVE
177	COMPLIANCE	227	LIQUID	277	PARTICULATE
178	AERATION	228	REACTIONS	278	INFILTRATION
179	ALUMINUM	229	SBRs	279	MINIMIZE
180	SCRAP	230	TESTS	280	RATES
181	ROCK	231	REACTION	281	SLURRY
182	IMPACTS	232	BLENDS	282	SODIUM
183	RESIDUAL	233	PIPE	283	SIGNIFICANT
184	COLLECTION	234	REGULATIONS	284	PACKAGING
185	GROUND	235	CAPACITY	285	REQUIRE
186	SULFATE	236	BASED	286	FUEL
187	FILTERS	237	WETLAND	287	LINER

188	MIXED	238	SEPARATION	288	EMISSION
189	CHLORINATED	239	PROPERTIES	289	SUSTAINABILITY
190	TOTAL	240	ODORS	290	BIOREACTOR
191	LABORATORY	241	APPLICATIONS	291	FLUID
192	FACILITIES	242	TABLE	292	PERMEABLE
193	DAM	243	PONDS	293	BACTERIA
194	CHEMICALS	244	IRON	294	PUMPED
195	PLANTS	245	WOOD	295	APPLICATION
196	ADDITIONAL	246	SOURCE	296	EROSION
197	SULFUR	247	MODELING	297	STORM
198	ION	248	AREA	298	DEWATERING
199	METHOD	249	MOISTURE	299	DIVERSION
200	REGULATORY	250	RECLAMATION	300	CONTAINMENT
301	CONDUCTIVITY	351	PRESSURE	401	OBSERVED
302	CLEAN	352	MIX	402	NUTRIENTS
303	REMOVED	353	SURFACTANT	403	ENGINEERING
304	EQUATION	354	MINERAL	404	RESIDUALS
305	INORGANIC	355	ENERGY	405	HYPOCHLORITE
306	BENEFICIATION	356	REDUCED	406	PEROXIDE
307	MOLDING	357	SOLVENT	407	COMPACTION
308	CSO	358	CONTAINER	408	SORPTION
309	OXIDANT	359	AMOUNT	409	METHANE
310	ASPHALT	360	LFG	410	STEAM
311	TOXIC	361	TYPICAL	411	LAYER
312	TOXICITY	362	STORMWATER	412	CHLORIDE
313	TANKS	363	MEDIA	413	FORMATION
314	SOLUBLE	364	DISINFECTANTS	414	INJECTED
315	PYRITE	365	PORTION	415	COLLOIDS
316	PRB	366	ASH	416	UNDERGROUND
317	MUNICIPALITIES	367	SLAG	417	ANALYSIS
318	STABILIZATION	368	GENERALLY	418	DIAMETER

319	DOWNSTREAM	369	MEASURED	419	BIODEGRADATION
320	BACKFILL	370	RESIN	420	STREAMS
321	AVERAGE	371	MAXIMUM	421	OCCUR
322	SAMPLE	372	LABOR	422	ASSOCIATED
323	CONSTRUCTED	373	ONSITE	423	PERFORMED
324	CONCRETE	374	FERROUS	424	DEPTH
325	NITRATE	375	RECOVERED	425	PROCESSED
326	MINERALS	376	PILES	426	PILOT
327	HYDROXIDE	377	PORE	427	CHLORITE
328	INSTALLATION	378	DRY	428	OXIDE
329	DISCHARGES	379	DISPERSION	429	EVALUATED
330	SOLVENTS	380	TRUCK	430	PCE
331	IMPOUNDMENTS	381	GENERATED	431	CLEANING
332	CARBONATE	382	BELOW	432	INCREASED
333	FIELD	383	APPENDIX	433	TRUCKS
334	PRIOR	384	NUTRIENT	434	EVALUATE
335	DIFFUSION	385	MINES	435	FRACTION
336	REDUCING	386	RESINS	436	EMBANKMENT
337	RATIO	387	SECONDARY	437	EVAPORATION
338	PERMIT	388	ZINC	438	COMPOSITE
339	PIT	389	POROUS	439	THERMAL
340	PRIMARY	390	SETTLING	440	DISPOSED
341	HYDROCARBONS	391	RECIRCULATION	441	SUBSTRATE
342	CALCIUM	392	DECREASE	442	SOURCES
343	VOLATILE	393	DETERMINED	443	TKN
344	ANALYZED	394	VALUES	444	BYPRODUCTS
345	MANAGEMENT	395	SULFURIC	445	OXIDIZING
346	VELOCITY	396	PRODUCTION	446	COARSE
347	ESTIMATED	397	MANUFACTURING	447	VOC
348	INCREASE	398	SOLUTIONS	448	MANUFACTURERS
349	DUST	399	WETLANDS	449	COMPACTED

350	VOCs	400	DESIGNED	450	GASES
451	BINDER	501	MECHANICAL	551	IMPACT
452	PVC	502	STRIPPING	552	WEIGHT
453	VERTICAL	503	BEHAVIOR	553	EFFICIENCIES
454	SWEEPINGS	504	KILN	554	OPERATORS
455	AMBIENT	505	CURBSIDE	555	VEGETATION
456	EXPOSURE	506	BEDROCK	556	ACIDIC
457	SILICA	507	PULP	557	BENEFICIAL
458	INLET	508	MOLD	558	ALKALINE
459	DISCHARGED	509	RANGE	559	MOLYBDENUM
460	DUMP	510	TCLP	560	REMEDIATE
461	ANALYTICAL	511	SIGNIFICANTLY	561	GASEOUS
462	DEBRIS	512	ALKALINITY	562	BIODEGRADABLE
463	PHOSPHORUS	513	CHLORAMINES		
464	MERCURY	514	DNAPL		
465	TESTED	515	DEPENDING		
466	OVERFLOW	516	CONSTITUENTS		
467	FOUNDRIES	517	REQUIRES		
468	REACTORS	518	RECYCLABLES		
469	FLUSHING	519	GRANULAR		
470	SPECIES	520	BOTTLES		
471	GEOCHEMICAL	521	HYDROSEEDING		
472	CHROMIUM	522	CEMENT		
473	TYPES	523	STANDARDS		
474	RELATIVELY	524	MANGANESE		
475	BASINS	525	COMPONENTS		
476	BARRIER	526	BLENDING		
477	FERRIC	527	GYPSUM		
478	ZONES	528	COLUMN		
479	CHLORINATION	529	APPLIED		
480	SEWAGE	530	MOLDED		

481	HYDROGEN	531	LOCATIONS		
482	SEDIMENTATION	532	NAPL		
483	GRAVITY	533	UPSTREAM		
484	RESULTING	534	REDOX		
485	STRENGTH	535	PETROLEUM		
486	SEVERAL	536	OPERATOR		
487	SOLUBILITY	537	WASTEWATERS		
488	DRINKING	538	OPERATIONAL		
489	INDUSTRY	539	COMMONLY		
490	PROVIDE	540	REAGENT		
491	SPECIFICATIONS	541	SLIMES		
492	OIL	542	AMD		
493	THERMOPLASTIC	543	MONOCHLORAMINE		
494	OZONATION	544	TREAT		
495	IONS	545	SLOPE		
496	DUMPS	546	TENSILE		
497	POTASSIUM	547	LINERS		
498	NATURAL	548	EQUILIBRIUM		
499	CONVENTIONAL	549	COVER		
500	RECYCLE	550	OCCURS		

2. Subcorpora Lists

This classification of the 500 analysed keywords is based on the detailed consistency analysis of the subcorpora. According to this analysis, some words occur more frequently in one or two subcorpora whereas others are spread more evenly throughout the whole corpus. In this appendix, the ones belonging to the subcorpora are listed first; the last batch of words are those with even distribution.

Subcorpus of webjournal texts

BENEFICIATION	EMISSIONS	POLLUTION
BIOREACTOR	LANDFILL	RECYCLING
COMPACTION	LANDFILLS	STORMWATER
COMPLIANCE	LEACHATE	TRUCK
COMPOSTING	LFG	TRUCKS
DIAMETER	METHANE	VOCS
DIVERSION	MUNICIPAL	
EMISSION	MUNICIPALITIES	

Subcorpus of mining texts

ACID	INFILTRATION	RECLAMATION
BACKFILL	LEACH	REMEDIAL
DAM	LEACHING	RUNOFF
DISCHARGE	LINER	SEEPAGE
DISCHARGES	ORE	SLAG
DOWNSTREAM	OXIDE	SLURRY
DRAINAGE	PIT	SULFATE
EMBANKMENT	POND	SULFIDE
EVAPORATION	PONDS	SULFURIC
FLOTATION	PORE	TAILINGS
IMPOUNDMENT	PRECIPITATION	UNDERGROUND
IMPOUNDMENTS	PYRITE	URANIUM

Subcorpus of remediation texts

ADSORPTION	DISINFECTANT	PERMEABLE
AQUEOUS	DISINFECTION	PEROXIDE
AQUIFER	DISSOLVED	PH
BACTERIA	DOWNGRADIENT	PHYTOREMEDIATION

BIODEGRADATION	GROUNDWATER	PLUME
BIOREMEDIATION	HYDRAULIC	POROUS
BYPRODUCTS	HYDROCARBONS	PRB
CHLORINATED	HYPOCHLORITE	REMEDIATION
CHLORINE	INACTIVATION	RESIDUAL
CHLORITE	INJECTED	SEDIMENT
CLEANUP	INJECTION	SEDIMENTS
COLLOID	INORGANIC	SITU
COLLOIDS	ION	SODIUM
CONCENTRATIONS	LABORATORY	SOLVENTS
CONDUCTIVITY	ORGANIC	SORPTION
CONTAMINANT	OXIDANT	TCE
CONTAMINANTS	OXIDATION	TECHNOLOGIES
CONTAMINATED	OZONE	TRACER
CONTAMINATION	PCE	UV
DEGRADATION	PERMANGANATE	ZONE
DIOXIDE	PERMEABILITY	

Subcorpus of technology fact sheets

ACTIVATED	FEED	SBRS
AERATION	FILTER	SEPTIC
AMMONIA	FILTERS	SEWER
ANAEROBIC	GAC	SLUDGE
ANOXIC	HYDRAULIC	SOLUBLE
BIOLOGICAL	INFLUENT	STORAGE
BIOMASS	NITRATE	TANK
BIOSOLIDS	NITRIFICATION	TANKS
BOD	NITROGEN	TOXICITY
CARBON	NUTRIENT	UV
COD	ODOR	WASTEWATER
DEGRADATION	ODORS	WETLAND

DEWATERING	OXYGEN	WETLANDS
DISINFECTION	RESIDUALS	
EFFLUENT	SBR	

Textbook subcorpus

ACTIVATED	EMISSIONS	POLLUTANT
AQUEOUS	FLUID	POLLUTANTS
BOD	FRACTION	POLLUTION
CARBON	FUEL	RESIN
COMBUSTION	HAZARDOUS	SLUDGE
CONTAMINANT	HYDROXIDE	SORPTION
DENSITY	MICROBIAL	SULFUR
DIAMETER	MICROORGANISMS	VAPOR
DIFFUSION	ORGANIC	VELOCITY
DISPERSION	OXIDIZED	VOLATILE
DISSOLVED	OXYGEN	
EMISSION	PARTICULATE	

Subcorpus of waste and recycling texts

ASPHALT	FEEDSTOCK	MSW
BATCH	FIBER	PACKAGING
BLENDS	FIBERS	RECYCLED
CALCIUM	FOUNDRY	RECYCLING
CARBONATE	HDPE	RESINS
COMPOST	LIME	REUSE
COMPOSTING	MOISTURE	SCRAP
CONCRETE	MOLDING	SWEEPINGS

Words which are evenly distributed throughout the corpus

ADDITION	GAS	PUMPED
ADDITIONAL	GENERALLY	PUMPING
AIR	GENERATED	QUALITY
ALUMINUM	GENERATION	RATE
AMOUNT	GLASS	RATIO
ANALYSIS	GROUND	REACTION
ANALYZED	IMPACTS	REACTIVE
APPENDIX	INCLUDE	REACTOR
APPLICATION	INCREASE	RECIRCULATION
APPROXIMATELY	INSTALLATION	RECOVERED
AREA	INSTALLED	RECOVERY
ASH	IRON	REDUCE
ASSOCIATED	LABOR	REDUCED
AVERAGE	LEVELS	REDUCING
BASED	LIQUID	REDUCTION
BASIN	LOADING	REGULATIONS
BELOW	LOCATED	REGULATORY
BINDER	MAINTENANCE	REMOVAL
CAPACITY	MANAGEMENT	REMOVE
CHARACTERISTICS	MANUFACTURING	REMOVED
CHARACTERIZATION	MASS	REQUIRE
CHEMICAL	MATERIAL	REQUIRED
CHEMICALS	MATERIALS	REQUIREMENTS
CHLORIDE	MAXIMUM	RESULTS
CLAY	MEASURED	ROCK
CLEAN	MEDIA	RUBBER
COLLECTED	METAL	SAMPLE
COLLECTION	METALS	SAMPLING
COLOR	METHOD	SAND
COMPOSITE	MILL	SECONDARY

COMPOUNDS	MINE	SEPARATION
CONCENTRATION	MINERAL	SETTLING
CONDITIONS	MINERALS	SIGNIFICANT
CONDUCTED	MINES	SITE
CONSTRUCTED	MINIMIZE	SITES
CONSTRUCTION	MINING	SITE-SPECIFIC
CONTAINER	MIX	SIZE
CONTAINMENT	MIXED	SOIL
CONTENT	MIXING	SOLID
CONTROL	MODELING	SOLIDS
COPPER	MONITORING	SOLUTION
CSO	NUTRIENTS	SOLVENT
DATA	OBSERVED	SOURCE
DECREASE	OCCUR	SPECIFIC
DENITRIFICATION	ONSITE	STABILIZATION
DEPTH	ON-SITE	STEAM
DESIGN	OPERATING	STORM
DESIGNED	OPERATION	STREAM
DETERMINE	ORGANICS	STREAMS
DETERMINED	PARAMETERS	STUDY
DISPOSAL	PARTICLES	SURFACE
DISPOSED	PER	SUSTAINABILITY
DUST	PERCENT	TABLE
EFFECTIVE	PERFORMANCE	TECHNOLOGY
EFFICIENCY	PERFORMED	TEMPERATURE
ENERGY	PERMIT	TEST
ENGINEERING	PHASE	TESTING
ENVIRONMENTAL	PILES	TIRE
EQUATION	PIPE	TOTAL
EQUIPMENT	PLANT	TOXIC
EROSION	PORTION	TRANSPORT
ESTIMATED	POTENTIAL	TREATMENT

EVALUATE	PRIMARY	TYPICAL
EVALUATED	PRIOR	TYPICALLY
EXPOSURE	PROCESS	VALUES
EXTRACTION	PROCESSED	VOLUME
FACILITIES	PROCESSES	WASTE
FACILITY	PROCESSING	WATER
FERROUS	PRODUCT	WELLS
FIELD	PRODUCTION	WOOD
FIGURE	PROGRAMS	ZINC
FILTRATION	PROJECT	
FLOW	PROJECTS	
FLOWS	PROPERTIES	
FORMATION	PUMP	

Appendix B – Sample Usage Profiles

The conventions of the usage profiles are described in 4.2. At this point, I just want to mention one more point: whenever verbs are written in capital print, they stand for the whole lemma of the verb.

pH (1600x in corpus) abbreviation rank 30 in keyword list	
N-1	**N+1**
equilibrium pH (13)	pH adjustment(s) (14)
optimum pH (12)	pH buffer(s) (6)
paste pH (6)	pH change(s) (6)
soil pH (10)	pH control(s) (9)
water pH (6)	pH gradient (5)
acidic pH (13)	pH increase (16)
alkaline pH (6)	pH level(s) (26)
ambient pH (11)	pH measurement(s) (6)
elevated pH (14)	pH range(s) (28)
high/er pH (58)	pH value(s) (61)
low/er pH (54)	pH dependent (6)
neutral pH (23)	pH DROP (6)
optimal pH (12)	pH DECREASE (11)
	pH INCREASE (11)
	pH RANGE (5)
clusters	
at a pH of (19) 5.7/9.5/…	
with a pH of (9) 5.3/9.2/…	
at pH (50) 1.8/5.6/…	
influenced by pH (5)	
BE analysed for pH (5)	
high/er pH level(s) (6)	
high/er pH values (15)	
change(s) in pH (11)	
decrease(s) in pH (12)	
gradient(s) in pH (5)	
increase(s) in pH (19)	
with increasing pH (16)	
(at) low/er pH values (13)	
at low pH (17)	
low/er pH levels (6)	
as a function of pH (13)	
ADJUST the pH (5)	
DETERMINE the pH (6)	
INCREASE the pH (12)	
lower the pH (7)	
MAINTAIN the pH (8)	
RAISE the pH (25)	

REDUCE the pH (10)
low/er/high/er/extreme/varying/… pH conditions (29)

VOC(s) (Volatile Organic Compound(s)) (335x in corpus) abbreviation rank 394/507 in keyword list	
N-1	**N+1**
chlorinated VOCs (9)	VOC concentration(s) (31)
total VOCs (5)	VOC emissions (15)
	VOC level(s) (5)
	VOC removal (5)
clusters	
remove VOCs/volatile organic compounds	
VOC from (9) (absorbents/an aquifer/groundwater/painting/ the offgas/…)	
VOC(s) in (19) (control samples/ groundwater/soil samples/the paint/the influent/the water/…)	

corpus-attested word-formation patterns	
derivational	**compounding**
semi-VOCs (1)	high-VOC (content) (1)
	SVOC(s) (6)
	CVOC (3)
	VOCcontaining (soil vapors) (1)
	VOC-content (exhaust streams/paints) (2)
	VOC-emitting (equipment)(1)
	VOC-free ((potable) water) (2)
	VOC-laden (wastewater/ airstreams/fumes) (4)
	VOC-treatment (technology) (1)

acidic (175x in corpus) adjective rank 635 in keyword list	
N-1	**N+1**
highly acidic (8)	acidic conditions (13)
slightly acidic (10)	acidic discharge(s) (5)
	acidic drainage (9)
	acidic pH (8)
	acidic solution(s) (10)
	acidic water(s) (9)
clusters	
under acidic conditions (8)	

corpus-attested word-formation patterns	
derivational	**compounding**
acid(s) (1687)	
acidity/ies (42)	
acidification (20)	
acidify (1)	
acidified (18)	

acidifies (2)	
acidifying (1)	

permeable (246 x in corpus) adjective rank 329 in keyword list	
N-1	**N+1**
highly permeable (15)	permeable barrier(s) (16)
less permeable (14)	permeable layer(s) (16)
more permeable (20)	permeable medium/media (8)
	permeable soil(s) (10)
	permeable zone(s) (15)
clusters	
slowly permeable (5) soil(s)/soil layer/shallow soil	
permeable reactive barrier wall(s) (5)	
permeable reactive barrier(s) (51)	
permeable reactive subsurface barrier(s) (7)	
permeable treatment zone(s) (8)	

corpus-attested word-formation patterns	
derivational	**compounding**
permeability/ies (365)	air-permeable material (1)
impermeable (92)	highly-permeable granular material (1)
impermeability (3)	low-permeable (2)
permeate (16)	vapor-permeable tubes (1)
permeates (1)	
permeating (2)	
permeation (4)	
non(-)permeable (3)	
semi(-)permeable (4)	

determined (801x in corpus) participial adjective/past participle rank 464 in keyword list	
N-1	**N+1**
experimentally determined (8)	
clusters	
determined using (21) (a concentrated strong acid/GIS/samples/site-specific materials/the short shot method/…)	
determined on a case(-)by(-)case basis (5)	
determined by (181) (a number of factors/analysis of …/analysing/bench testing/computing/monitoring/multiplying/…)	
determined for (54) (a given situation/each pathway/the site/the sample/various treatment trains/…)	
determined through (8) (air-quality monitoring/bench studies/experimentation/…)	

corpus-attested word-formation patterns	
derivational	**compounding**
determination(s) (199)	
determine (1071)	

determines (69)	
determining (204)	
determinative (2)	
determinant(s) (4)	
indeterminate (1)	
predetermine (2)	
predetermined (22)	
pre-determined (6)	
undetermined (7)	
selfdetermination (1)	

recovered (480x in corpus) participial adjective/past participle rank 427 in keyword list	
N-1	N+1
	recovered material(s) (24)
	recovered metal(s) (5)
	recovered paper (23)
	recovered product(s) (6)
	recovered vapor(s) (6)
	recovered wastepaper (28)
clusters	
recovered at a rate of (11) 98 per cent/16.6 percent/40 percent/…	
recovered for recycling (23)	
recovered for reuse (5)	
recovered from (46) batteries/mines/solution/the brine/the whole tailings/wells/…	
recovered through (8) a collection program/curbside programs/the recycling program/…	

corpus-attested word-formation patterns	
derivational	**compounding**
recover (147)	
recovers (18)	
recovering (33)	
recovery/ies (1147)	
recoverable (24)	
recoverability (1)	

on-site/onsite/on site (507 x in corpus) adverb/adjective rank 418 in keyword list	
N-1	**N+1**
generated on(-)site (16)	on(-)site disposal (5)
stored on(-)site (17)	on(-)site generation (5)
	on(-)site regeneration (5)
	on(-)site system(s) (8)
	on(-)site treatment (10)
	on(-)site use(s) (7)
clusters	
on(-)site wastewater treatment (9)	

downgradient (231x in corpus) adj./adverb rank 297 in keyword list	
N-1	**N+1**
distance downgradient (8)	downgradient pumping (5)
MOVE downgradient (6)	downgradient well(s) (7)
clusters	
on the downgradient side (9)	
downgradient from (35) (the disposal beds/ injection location/ponds/site/test zone/…)	
downgradient monitoring well(s) (8)	
downgradient of (the barrier/the dam/the hot spot/the injection line/the PRB/the wall/…)	

significantly (526x in corpus) adverb rank 586 in keyword list	
N-1	**N+1**
CHANGE significantly (10)	significantly different (20)
CONTRIBUTE significantly (14)	significantly IMPROVE (10)
DECREASE significantly (6)	
DROP significantly (7)	
INCREASE significantly (16)	
REDUCE significantly (5)	
VARY significantly (25)	
clusters	
can/will/DO not/BE (not) significantly AFFECT (29)	
can/could/DO not significantly alter (5)	
significantly greater (than) (16)	
significantly higher (than) (31)	
BE significantly impacted (5)	
can/would/may/will/to significantly INCREASE (21)	
can/BE significantly INFLUENCE (6)	
significantly less (than) (24)	
significantly lower (than) (25)	
significantly more (19) companies/complex/variable/…	
can/BE/to significantly REDUCE (66)	

corpus-attested word-formation patterns	
derivational	**compounding**
significant (1177)	
significance (53)	

biodegradation (164x in corpus) noun rank 471 in keyword list	
N-1	**N+1**
aerobic biodegradation (11)	biodegradation process(es) (13)
anaerobic biodegradation (6)	biodegradation rate(s) (7)
natural biodegradation (10)	
enhance biodegradation (7)	
stimulate biodegradation (5)	
clusters	
biodegradation of (34) (alcohol/benzene/contaminants/landfills/organic waste/PCBs/these chemicals/…)	

corpus-attested word-formation patterns	
derivational	**compounding**
biodegradability (19)	
biodegradable (135)	
biodegradables (3)	
biodegradative (1)	
biodegrade (17)	
biodegraded (16)	
biodegrades (1)	
biodegrading (6)	
non(-)biodegradable (16)	
non-biodegradables (1)	

infiltration (307x in corpus) noun rank 309 in keyword list	
N-1	**N+1**
ground(-)water infiltration (9)	infiltration device(s) (5)
precipitation/rainfall/rainwater/snowmelt infiltration (6)	infiltration gallery/ies (14)
water infiltration (7)	infiltration problem(s) (5)
net infiltration (34)	infiltration rate(s) (20)
PREVENT infiltration (7)	infiltration sumps (8)
PROMOTE infiltration (7)	infiltration trenches (5)
REDUCE infiltration (9)	
clusters	
infiltration into (22)(the tanks/the tailings/the source/(the) mine rock piles (5)…)	
infiltration through (7) (the tailings/the swale/dump materials/the landfill/the barrier layer/…)	
relative infiltration tests/rates (6)	
storm water infiltration sumps (8)	
rate(s) of net infiltration (14)	

corpus-attested word-formation patterns	
derivational	**compounding**
infiltrate (14)	infiltration-limiting (1)
infiltrated (5)	
infiltrating (15)	
infiltrative (5)	
reinfiltrating (1)	
re-infiltration (1)	

particle(s) (1578x in corpus) noun rank 135/107 in keyword list	
N-1	**N+1**
algae particle(s) (5)	particle capture (7)
ash particle(s) (9)	particle concentration (13)
clay particle(s) (13)	particle density (9)
dirt particles (5)	particle deposition (12)

dust particles (9)	particle diameter(s) (11)
glass particle(s) (31)	particle emissions (6)
iron particle(s) (11)	particle level(s) (6)
rubber particles (25)	particle mass (9)
sand particles (7)	particle removal (21)
smoke particles (7)	particle settling (10)
soil particles (30)	particle shape(s) (10)
soot particles (9)	particle size (172)
titania particles (6)	particle surface (10)
wood particles) (14)	particle tracking (5)
airborne particle(s) (10)	particles PASS (6)
charged particles (11)	particle(s) SETTLE (7)
coarse particles (20)	
colloidal particles (27)	
filtered particles (5)	
fine particles (50)	
large/r particle(s) (37)	
small/er/est particle(s) (61)	
solid particles (13)	
spherical particles (14)	
suspended particles (25)	
turbid particles (6)	
REMOVE particles (11)	
clusters	
average particle size (5)	
adhesion/bonding/bonds/collisions/relationships/repulsion between particles (9)	
wide/full/broad/moderate/total range of particle sizes (5)	
rubber particle size (25)	
rubber particle surface (7)	
REDUCE the particle size (5)	
particles are removed (6)	
particle deposition rate(s) (9)	
particles in water (5)	
particle size distribution (22)	
particle size reduction (6)	
particles suspended in (10) a fluid/a liquid /a solution	

corpus-attested word-formation patterns	
derivational	**compounding**
interparticle (5) collisions/contact/bridging	gross alpha-particle activity (1)
	coarse-particle (2) fraction/removal
	fine-particle (5) fraction/removal/ sand-stone/...
	gas-to-particle conversion process(es) (2)
	one-particle-thick (2) stream/feed
	particle-size (4) dependence/gradation/ distribution
	particle-sized carbon blacks (2)
	particle-to-particle collisions (1)

	particle-tracking software (1)
	particleboard (2)
recycling (1673x in corpus) verbal noun/participial adjective rank 20 in keyword list	
N-1	**N+1**
aerosol recycling (6)	recycling authorities (6)
biosolids recycling (8)	recycling business(es) (7)
container recycling (11)	recycling center(s) (15)
curbside recycling (5)	recycling company/ies (15)
electronic(s) recycling (11)	recycling coordinator(s) (15)
feedstock recycling (7)	recycling cost(s) (9)
material(s) recycling (8)	recycling depot(s) (5)
packaging recycling (8)	recycling effort(s) (14)
paper recycling (12)	recycling equipment (12)
plastic(s) recycling (22)	recycling facility/ies (43)
product recycling (19)	recycling firm(s) (5)
textile recycling (30)	recycling goal(s) (5)
tire recycling (13)	recycling industry/ies (25)
waste recycling (20)	recycling infrastructure (5)
mechanical recycling (7)	recycling initiative(s) (6)
increase recycling (5)	recycling markets (11)
	recycling operation(s) (14)
	recycling opportunities (8)
	recycling option(s) (16)
	recycling plant(s) (15)
	recycling process(es) (24)
	recycling program(s) (79)
	recycling rate(s) (26)
	recycling services (8)
	recycling system(s) (23)
	recycling truck(s) (7)
clusters	
curbside recycling program(s) (11)	
material(s) for recycling (8)	
recovered for recycling (23)	
recovery for recycling (35)	
single-stream recycling (5)	
scrap textile recycling (19)	
scrap tire recycling (5)	

corpus-attested word-formation patterns	
derivational	**compounding**
recycle (189)	computer-recycling program (1)
recycled (1196)	curbside-recycling program (1)
recycles (21)	
recyclable (101)	
recyclables (126)	
recyclability (5)	

non-recycling (1)	
post-recycling (2) process/residential waste	

sampling (774x in corpus) verbal noun/participial adjective rank 113 in keyword list	
N-1	**N+1**
gas sampling (6)	sampling activities (7)
ground(-)water sampling (19)	sampling approach(es) (5)
performance sampling (7)	sampling data (10)
soil sampling (9)	sampling design(s) (40)
water sampling (10)	sampling device(s) (15)
additional sampling (6)	sampling equipment (10)
previous sampling (5)	sampling event(s) (28)
spatial sampling (7)	sampling location(s) (6)
statistical sampling (10)	sampling method(s) (9)
synoptic sampling (26)	sampling objective(s) (5)
quarterly sampling (8)	sampling period(s) (8)
	sampling plan(s) (10)
	sampling point(s) (17)
	sampling port(s) (29)
	sampling procedure(s) (12)
	sampling program(s) (30)
	sampling results(s) (16)
	sampling round(s) (8)
	sampling strategy/ies (8)
	sampling study/ies (7)
	sampling technique(s) (8)
clusters	
compliance sampling program (5)	
prior to sampling (6)	
sampling design strategy/ies (7)	

corpus-attested word-formation patterns	
derivational	**compounding**
sample(s) (2276)	batch-sampling CEMS (10)
sampler(s) (65)	direct-sampling device (2)
sampled (79)	gas-sampling (3) tubing/tubes/port depth
pre-sampling activities (1)	soil-sampling boundary (1)
presampling (5)	

seepage (410 x in corpus) noun rank 161 in keyword list	
N-1	**N+1**
impoundment seepage (7)	seepage collection (9)
tailings seepage (54)	seepage control(s) (13)
contaminated seepage (5)	seepage flow(s) (13)
prevent seepage (6)	seepage model (5)
reduce seepage (5)	seepage modelling (5)

	seepage pond(s) (5)
	seepage quality (5)
	seepage rates (9)
	seepage water (5)
clusters	
seepage from (54) (irrigation systems/mine waste/mine adits/tailings impoundments/underground mine workings/...)	
seepage return pond (7)	
seepage through (16) (a liner/embankments/the piles/the rock mass/dam/weak permeable layers/...)	

corpus-attested word-formation patterns	
derivational	**compounding**
seep (18)	
seeped (3)	
seeping (8)	
seeps (23)	

sorption (156 x in corpus) noun rank 460 in keyword list	
N-1	**N+1**
	sorption isotherm(s) (8)
	sorption site(s) (6)
clusters	
activated carbon sorption (5)	
sorption of (18) ((the) contaminant(s)/carbofuran/colloidal COD/organics/organic molecules/PCBs and surfactants/...)	
sorption on (9) ((a grain of/granular/powered) activated carbon/solids/soil grains/...)	

corpus-attested word-formation patterns	
derivational	**compounding**
sorptive (5)	sorption-based (3) (processes/solidifaction)
absorption (254)	
absorptive (2)	
adsorption (329)	
adsorptive (12)	
adsorptivity (1)	
desorption (98)	

determine (1071x in corpus) verb rank 138 in keyword list
clusters
determine if … (104)
determine whether … (84)
determine the concentration(s) of … (12)
determine the effect of … (17)
determine the effectiveness of … (9)
determine the efficiency (5)
determine the extent of … (7)
determine the feasibility of …(22)
determine (the) isotherm parameters (10)
determine the level of … (5)
determine the pH (of …) (5)
determine the rate of … (5)
determine the size (of …) (7)
determine the suitability of … (5)
determine the value(s) of … (5)
determine the volume of … (7)

corpus-attested word-formation patterns	
derivational	compounding
determination(s) (199)	
determinative (2)	
determinant(s) (4)	
indeterminate (1)	
predetermine (2)	
predetermined (22)	
pre-determined (6)	
undetermined (7)	

minimize/se (341x in corpus) verb rank 311 in keyword list	
N-1	N+1
	minimize waste(s) (7)
clusters	
minimize the amount of (9) (organic substrate/material/surfactants/surface water/…)	
minimize (the) costs (of…) (6)	
minimize the effect(s) of … (7)	
minimize the (environmental) impact(s) of ... (7)	
minimize the potential for .. (7)	
minimize (the) risks (of …) (13)	

corpus-attested word-formation patterns	
derivational	compounding
minimization (103)/minimisation (2)	

FSC
www.fsc.org
MIX
Papier | Fördert gute Waldnutzung
FSC® C083411

Kolibri 360 GmbH
Ferdinand-Jühlke-Straße 7
99095 Erfurt, Deutschland
produktsicherheit@kolibri360.de

Druck:
CPI Druckdienstleistungen GmbH
im Auftrag der
Kolibri 360 GmbH
Ferdinand-Jühlke-Str. 7
99095 Erfurt